Highways to Success
or Byways to Waste

Highways to Success or Byways to Waste

Estimating the Economic Benefits of Roads in Africa

Rubaba Ali, A. Federico Barra, Claudia Berg,
Richard Damania, John Nash, and Jason Russ

A copublication of Agence Française de Développement and the World Bank

Africa Development Forum Series

The **Africa Development Forum Series** was created in 2009 to focus on issues of significant relevance to Sub-Saharan Africa's social and economic development. Its aim is both to record the state of the art on a specific topic and to contribute to ongoing local, regional, and global policy debates. It is designed specifically to provide practitioners, scholars, and students with the most up-to-date research results while highlighting the promise, challenges, and opportunities that exist on the continent.

The series is sponsored by the Agence Française de Développement and the World Bank. The manuscripts chosen for publication represent the highest quality in each institution and have been selected for their relevance to the development agenda. Working together with a shared sense of mission and interdisciplinary purpose, the two institutions are committed to a common search for new insights and new ways of analyzing the development realities of the Sub-Saharan Africa region.

Advisory Committee Members

Agence Française de Développement
Jean-Yves Grosclaude, Director of Strategy
Alain Henry, Director of Research
Guillaume de Saint Phalle, Head of Research and Publishing Division
Cyrille Bellier, Head of the Economic and Social Research Unit

World Bank
Francisco H. G. Ferreira, Chief Economist, Africa Region
Stephen McGroarty, Executive Editor, Publishing and Knowledge Division
Carlos Rossel, Publisher

Sub-Saharan Africa

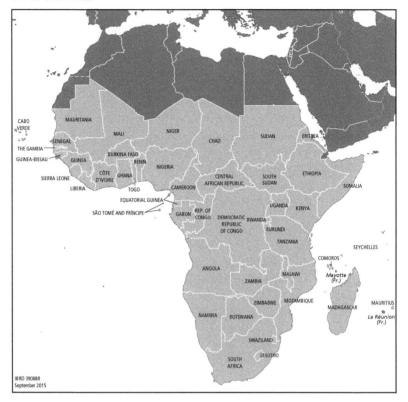

Source: World Bank (IBRD 39088R, September 2015).

Titles in the Africa Development Forum Series

Africa's Infrastructure: A Time for Transformation (2010) edited by Vivien Foster and Cecilia Briceño-Garmendia

Gender Disparities in Africa's Labor Market (2010) edited by Jorge Saba Arbache, Alexandre Kolev, and Ewa Filipiak

Challenges for African Agriculture (2010) edited by Jean-Claude Deveze

Contemporary Migration to South Africa: A Regional Development Issue (2011) edited by Aurelia Segatti and Loren Landau

Light Manufacturing in Africa: Targeted Policies to Enhance Private Investment and Create Jobs (2012) by Hinh T. Dinh, Vincent Palmade, Vandana Chandra, and Frances Cossar

Informal Sector in Francophone Africa: Firm Size, Productivity, and Institutions (2012) by Nancy Benjamin and Ahmadou Aly Mbaye

Financing Africa's Cities: The Imperative of Local Investment (2012) by Thierry Paulais

Structural Transformation and Rural Change Revisited: Challenges for Late Developing Countries in a Globalizing World (2012) by Bruno Losch, Sandrine Fréguin-Gresh, and Eric Thomas White

The Political Economy of Decentralization in Sub-Saharan Africa: A New Implementation Model (2013) edited by Bernard Dafflon and Thierry Madiès

Empowering Women: Legal Rights and Economic Opportunities in Africa (2013) by Mary Hallward-Driemeier and Tazeen Hasan

Enterprising Women: Expanding Economic Opportunities in Africa (2013) by Mary Hallward-Driemeier

Urban Labor Markets in Sub-Saharan Africa (2013) edited by Philippe De Vreyer and François Roubaud

Securing Africa's Land for Shared Prosperity: A Program to Scale Up Reforms and Investments (2013) by Frank F. K. Byamugisha

Youth Employment in Sub-Saharan Africa (2014) by Deon Filmer and Louis Fox

Tourism in Africa: Harnessing Tourism for Growth and Improved Livelihoods (2014) by Iain Christie, Eneida Fernandes, Hannah Messerli, and Louise Twining-Ward

Safety Nets in Africa: Effective Mechanisms to Reach the Poor and Most Vulnerable (2015) edited by Carlo del Ninno and Bradford Mills

Land Delivery Systems in West African Cities: The Example of Bamako, Mali (2015) by Alain Durand-Lasserve, Maÿlis Durand-Lasserve, and Harris Selod

Enhancing the Climate Resilience of Africa's Infrastructure: The Power and Water Sectors (2015) edited by Raffaello Cervigni, Rikard Liden, James E. Neumann, and Kenneth M. Strzepek

Africa's Demographic Transition: Dividend or Disaster? (2015) edited by David Canning, Sangeeta Raja, and Abdo S. Yazbeck

The Challenge of Fragility and Security in West Africa (2015) by Alexandre Marc, Neelam Verjee, and Stephen Mogaka

Highways to Success or Byways to Waste: Estimating the Economic Benefits of Roads in Africa (2015) by Rubaba Ali, A. Federico Barra, Claudia Berg, Richard Damania, John Nash, and Jason Russ

All books in the Africa Development Forum series are available for free at https://openknowledge.worldbank.org/handle/10986/2150

Contents

Index   181

Boxes

Figures

Maps

Tables

Foreword

Roads are a vital driver of economic development. They connect people to jobs, schools, markets, and hospitals. Unfortunately, Africa remains the least connected region in the world, particularly in rural areas, where 70 percent of the poor live. Barely a third of rural Africans live within 2 kilometers of an all-season road, compared to some 65 percent in other developing regions. Lack of rural road connectivity constrains agricultural production and marketing, making it difficult and costly to get inputs in and products out.

While the benefits of improved roads are well recognized, it is not always clear where investments in roads will do the most good. A strategic road improvement program may be all it takes to boost local trade and spur economic growth, but poorly conceived investments can prove to be counterproductive.

Going forward, it is critical that we make use of the best tools available to prioritize transport projects and invest wisely. This report makes an important methodological contribution toward that end. Using spatial analytic tools, it develops an innovative approach to estimate the economic benefits of infrastructure, and demonstrates through simulations of alternative road investments which investments would yield the highest economic growth and which would benefit the greatest number of people. This investigation includes exploration of a question that has been neglected in previous research, but is of vital importance: When—and under what circumstances—does it make sense to invest in roads in fragile and conflict-prone areas? This report explores the relationship between conflict and the benefits derived from reducing transport costs in these regions.

The timing of the report is especially apt. While African economies are cooling off to some extent as the commodity boom subsides, prospects for continued income growth remain good. Africa could feed itself and create a trillion-dollar regional food market by 2030. However, in order to trade their products, farmers need better access to roads. Improved farm-to-market connectivity should be a priority, but one that will likely require significant investments.

Nervous investor sentiment combined with lower global prices for minerals and hydrocarbons will mean reduced revenues in government coffers. In that context, establishing the links between infrastructure investments and economic growth, and screening investments for sustainable growth, will be even more critical.

Africa is rich in natural resources, including lush forests and vibrant biodiversity. All too often, roads have been built in environmentally sensitive areas, opening these regions up to unfettered deforestation. Evaluating and mitigating the potential environmental costs of improving roads is therefore essential. By examining the impact of road improvement on deforestation and loss of biodiversity in the Congo Basin, this study makes an important contribution to an otherwise neglected area of research.

As this book demonstrates, methodologies based on spatial econometric techniques are useful additions to the tool kit for assessing environmental impacts, thereby helping to ensure that roads are not located in areas where they will undermine the goal of sustainable development.

This study will contribute to expanding the frontier of knowledge in a field that is important in enhancing both the rate and sustainability of Africa's growth.

Makhtar Diop
Vice President, Africa Region
World Bank

Acknowledgments

This volume is part of the African Regional Studies Program, an initiative of the Africa Region Vice Presidency at the World Bank. This series of studies aims to combine high levels of analytical rigor and policy relevance, and to apply them to various topics important for the social and economic development of Sub-Saharan Africa. Quality control and oversight are provided by the Office of the Chief Economist of the Africa Region.

This report has been prepared by a research team led by Richard Damania and John Nash, comprising Rubaba Ali, A. Federico Barra, Claudia Berg, and Jason Russ. David Wheeler of the World Resources Institute made key contributions designing the biodiversity index and providing extensive knowledge of issues related to deforestation in chapter 4. The work was conducted under the guidance and supervision of Francisco Ferreira, then chief economist of the World Bank's Africa Region. The authors gratefully acknowledge his support throughout the process of researching and writing the report.

During the early stages of the analysis, the team received excellent and extensive advice from a number of World Bank colleagues. Uwe Deichmann provided valuable technical guidance on spatial econometrics issues; Mohammed Dalil Essakali, senior transport economist, gave much appreciated data advice; and, importantly, Alberto Nogales guided us on the HDM-4 methodology. During the analysis, the authors benefited greatly from the country-specific knowledge of Marie-Francoise Mary Nelly, country director for Nigeria, and Sateh Chafic El-Arnaout, program leader for Nigeria, and Jean-Christophe Carret during his tenure as Sustainable Development Network sector leader in the Democratic Republic of Congo.

Special thanks go to the peer reviewers of the report who provided detailed comments and helpful feedback: Harris Selod, Somik Lall, Atsushi Iimi, Don Larson, as well as two anonymous referees. Following their invaluable suggestions, the analysis and presentation of the report were much improved. The authors are very grateful as well to Marianne Fay, Grahame Dixie, and Shahe Emran, who read the manuscript and offered insightful comments. The team would also like to acknowledge the helpful discussions held throughout the research process with Zoubida Allaoua, acting vice president for Sustainable Development, and with George Washington University (GW) professors Stephen C. Smith, James E. Foster, Arun Malik, Ram Fishman, and Ben Williams. We also wish to thank the participants of the North East Universities Development Consortium conference on November 1, 2014, and the Development Lunch held at GW on March 27, 2015.

About the Authors

Richard Damania is the global lead economist in the World Bank's Water Practice. Prior to this he was the lead economist of the Africa Sustainable Development Department with responsibility for infrastructure, environment, and social issues. He has also served as lead economist in the South Asia and Latin America and Caribbean Regions of the World Bank. Before joining the World Bank he was at the University of Adelaide in Australia. He has held numerous advisory positions at government and international organizations and serves on the editorial board of several academic journals in natural resource economics.

John Nash has worked at the World Bank in several regions and is now lead economist in the Agriculture Global Practice in the Africa Region. He holds MSc and PhD degrees in economics from the University of Chicago and a BS degree in economics from Texas A&M University. Prior to joining the World Bank in 1986, he was an assistant professor at Texas A&M University, and an economic adviser to the chairman of the U.S. Federal Trade Commission. He is married with a son and a daughter. He loves to ski and scuba with his wife, Sarah.

Rubaba Ali is a PhD candidate in the Department of Agricultural and Resource Economics at the University of Maryland, College Park. She is currently employed at Fannie Mae as an economist, providing guidance on analysis related to credit loss mitigation. Her research interests include the impact of transport infrastructure on agricultural and nonagricultural sectors, access to credit, and overall welfare in developing countries. She conducts research on credit, energy, and transport-related issues in developing and developed countries. She received master's degrees in economics and agricultural and resource economics from the University of Maryland, and undergraduate degrees in mathematics and economics from Bard College.

Authors are listed on the title page in alphabetical order.

A. Federico Barra is a land administration/geospatial specialist at the World Bank. Since joining the World Bank in 2008, he has been involved in numerous geospatial/economics knowledge and analytical products for the Africa Region, where he applied these innovative techniques in several sectors, including infrastructure; urban, rural, and social development; environment; and agriculture. Before joining the World Bank, he worked as a consultant at the Consultative Group to Assist the Poor (CGAP), Winrock International, and the United Nations Economic Commission for Latin America and the Caribbean (ECLAC). He holds a bachelor's degree in economics from Universidad Nacional de Córdoba (Argentina) and an MS degree in public policy and management from Carnegie Mellon University (Pittsburgh, United States).

Claudia Berg has been working at the World Bank since 2013, most recently joining the Development Research Group. During this time, her research has focused primarily on issues of rural development, in particular estimating the benefits of roads for rural producers in Africa. She previously worked at the Small Enterprise Assistance Funds (SEAF), where she evaluated the development impact of investments in small and medium enterprises. Prior to joining the World Bank, she worked at George Washington University where she taught microeconomics and earned a PhD degree in economics. Her research there focused on assessing the impact of microfinance on the ultrapoor in Bangladesh. She also holds a MSc degree in economics from the University of Essex in the United Kingdom and a BA degree in international economics from Franklin College (Switzerland).

Jason Russ is a candidate for the PhD degree in the economics department at George Washington University. He first joined the World Bank as a consultant in 2012 where he has been working on research related to the field of sustainable development and transport economics, utilizing a spatial econometric approach. Prior to joining the World Bank, he was a tax consultant at PricewaterhouseCoopers. He earned a BA degree from University of Maryland, College Park, and an MA degree in economics from Fordham University.

Abbreviations

ACLED	Armed Conflict Location Events Dataset
AICD	Africa Infrastructure Country Diagnostic
DHS	Demographic and Health Survey
DRC	Democratic Republic of Congo
FERMA	Nigeria Federal Roads Maintenance Agency
GDP	gross domestic product
GIS	geographic information system
GLS	generalized least squares
GLS-IV	generalized least squares with instrumental variables
HDM-4	Highway Development Management Model
IMR	inverse Mills ratio
ISA	Integrated Surveys on Agriculture
IUCN	International Union for Conservation of Nature
IV	instrumental variable
LSMS	Living Standards Measurement Study
MPI	multidimensional poverty index
NDHS	Nigeria Demographic and Healthy Survey
NEPAD	New Economic Partnership for Africa's Development
NHP	natural-historical path
OLS	ordinary least squares
REDD+	Reducing Emissions from Deforestation and Forest Degradation
SPAM	Spatial Production Allocation Model
2SLS	two-stage least squares
WWF	World Wildlife Fund

Chapter 1

Road Map to the Report

Introduction

This study develops and tests new approaches to the planning of infrastructure to maximize benefits and minimize negative externalities, particularly in rural areas. It explores several questions related to the impacts of infrastructure on welfare and poverty that are especially relevant for Sub-Saharan Africa. Reducing poverty in Sub-Saharan Africa—currently the poorest region in the world despite its widely acknowledged enormous potential for growth—is the world's supreme development challenge. GDP growth has picked up in recent years, but mostly driven by increased production of and higher prices for mineral and hydrocarbon resources. This growth model has not turned out to be an effective engine to drive poverty reduction or boost shared prosperity. Clearly one key to progress toward these goals is to kindle growth in rural areas, since that is where more than 70 percent of the continent's poorest populations live, with agriculture as their most important economic activity. The primary sector accounts for 15 percent of GDP regionwide and almost 60 percent of the labor force (Nash, Halewood, and Melhem 2013; World Bank 2014a), with agro-processing accounting for an additional 15–25 percent of GDP.

Further underscoring the need to encourage growth in rural areas, ongoing research using the Global Trade Analysis Project model of world trade finds that productivity growth in agriculture is almost three times as effective at reducing poverty as is growth in other sectors. A 1 percent improvement in agricultural productivity translates into about a 0.9 percentage point fall in poverty in developing countries, compared with a fall of 0.3–0.4 percentage point from a 1 percent increase in productivity in other sectors (Ivanic and Martin 2014). Agriculture is also critical for managing the urban transition that Africa will undergo. To date, this process has been driven to a large extent by populations being pushed out of rural areas, rather than by cities attracting a workforce by acting as growth poles. The urban transition would be a more positive process were it driven by improving economic opportunities in the cities that gradually pull rural residents in, rather than by declining conditions and periodic disasters in rural areas that push residents out. Such dislocations often create conflict and waves of migration that are difficult for cities to absorb, typically just leading to expanded slums. A key element of a transition strategy, therefore, is to enhance living conditions in rural areas, and this report argues that—with caveats and qualifications—improved transport linkages can make a significant contribution, and demonstrates a methodology for determining the magnitude and location of those benefits.

To set the stage for the rest of the report, this chapter first explains why there is good reason to believe that the prospects are bright for setting the agricultural sector in Africa on a high-growth trajectory, given proper conditions. It will show that one key element in achieving this shift will be improving farm-to-market connectivity, which is currently the worst of all regions in the world. However, given Sub-Saharan Africa's enormous needs and limited resources, investing wisely, that is, prioritizing potential investments, will be critically important. Therefore, tools are needed to accurately evaluate the benefits and costs of alternatives. This report seeks to demonstrate that local conditions matter considerably, and the presence or absence of conflict, environmental externalities, and local production potential are the focus of this investigation. Data and econometric issues pose formidable challenges to this effort. The chapter also briefly reviews the literature on estimating benefits to show how this report fits in. The introduction concludes with a short description of each of the subsequent chapters. In brief, chapters 2 and 3 examine the positive benefits of road investment on various measures of welfare, and chapters 4 and 5 look at the negative aspects.

The Untapped Potential of African Agriculture

With an abundance of labor, land, and water, Africa has the resources neces-sary for agricultural prosperity. Of the world's surface area suitable for sustain-able expansion of production—that is, nonprotected, unforested land, with low population density—Africa has the largest share by far, accounting for roughly 45 percent of the global total (Deininger and Byerlee 2011). Although some large areas of the continent are arid or semi-arid, available water resources in Africa are, on average, greatly underutilized. Only 2–3 percent of renewable water resources in Africa are being used, compared with 5 percent worldwide. Furthermore, relative to other regions, Africa has labor costs that are low relative to the cost of capital, which should encourage the production of labor-intensive farming-related products and services. One study found that minimum wages in Thailand, a major agricultural exporter, were 2–3 times costlier than those in Ghana and 1.6–2.2 times those in Senegal (Byerlee et al. 2013).

Despite these advantages, Africa has been steadily losing its share of both global and regional agricultural markets during the past 40 years. From a global share of 7–8 percent in the 1960s, Africa's agricultural exports fell to about

2 percent by the beginning of the 1990s. Today, Brazil has several times the share of the entire African continent, and even Thailand—with a tiny fraction of Africa's land area—has a larger market share (Byerlee et al. 2013).

A number of factors have contributed to the loss of competitiveness of Africa's agriculture, but certainly two important causes of the low productivity levels are the limited use of inputs and the slow adoption of improved production technologies. Africa has by far the lowest rate of fertilizer use of any region, a rate that has remained virtually the same since 1970 despite considerable efforts by governments and donors to raise it (Figure 1.1). Use of other yield-enhancing inputs—such as improved crop varieties, pesticides (herbicides, insecticides, fungicides), and mechanization—are similarly limited despite subsidized distribution by some governments. Moreover, participation in farmer extension and education programs aimed at increasing knowledge and skills meant to complement the use of higher-yielding inputs is low and has been declining (Pardey, Alston, and Piggot 2006). In the absence of proper management techniques, even low yields are not sustainable in the long term on currently cultivated lands because soils are being depleted of nutrients and, without fertilizer, they are not being replenished.

Figure 1.1 Use of Fertilizer in Africa Compared with Other Regions, 1970–2004

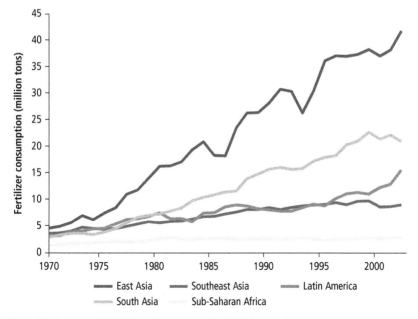

Source: Calculated from FAOSTAT (http://faostat.fao.org/site/339/default.aspx).

Figure 1.2 Aggregate Food Staple Trade by World Regions, 1990 and 2010

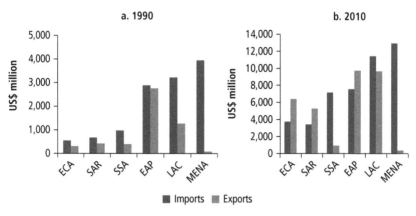

Source: COMTRADE, reported in World Bank 2012.
Note: ECA = Europe and Central Asia; SAR = South Asia; SSA = Sub-Saharan Africa; EAP = East Asia and the Pacific; LAC = Latin America and the Caribbean; MENA = Middle East and North Africa.

Regional market opportunities have also been missed, and the gap between regional demand and supply has widened, causing imports—particularly of food staples—to boom. From the 1990s to the 2000s, exports of food staples grew much faster than imports in Europe and Central Asia, South Asia, and East Asia and the Pacific; in Sub-Saharan Africa, however, imports skyrocketed to about 6.5 times their earlier level, while exports barely increased (Figure 1.2). Food trade deficits are understandable and even desirable in a region such as the Middle East and North Africa, which has no comparative advantage in food production. But in large parts of Sub-Saharan Africa, where all of the natural ingredients for efficient production are present, deficits of this nature signal that something fundamental is amiss.

If not reversed, the consequences of the missed opportunity to capture regional markets will grow over time as that market expands. If these markets could instead be captured by regional producers, the benefits would be enormous. Population and incomes are growing rapidly in the region, with a consequent increase in food demand, especially for higher-value products. In addition, Sub-Saharan Africa is undergoing the most rapid urbanization of any region, and urban consumers on average spend a significantly higher share of their incomes on food.[1] If the New Economic Partnership for Africa's Development's (NEPAD's) projections are correct, incomes will grow by 6 percent per year; assuming marginal expenditure on food of 0.5 percent (compared with the current figure of 0.6), food markets would reach US$1 trillion by 2030 (compared with the 2010 figure of US$313 billion), with much of the growth driven by urban demand (Figure 1.3).

Figure 1.3 Projected Increase in Value of Food Markets in Sub-Saharan Africa by 2030

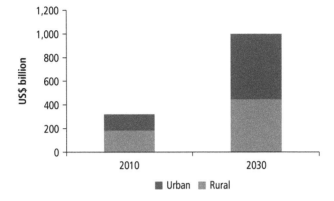

Source: Nash, Halewood, and Melhem 2013.

The Importance of Improving Connectivity

To take full advantage of this agricultural opportunity, existing barriers to market access and trade within and outside the region will need to be overcome. These barriers currently fragment natural "marketsheds," preventing the connection of food surplus and food deficit regions, reducing the welfare of both, and increasing reliance on external markets (World Bank 2012; Nash, Halewood, and Melhem 2013; Haggblade 2013).

However, integration is complicated by the region's economic geography. The spatial distribution of economic activity in Africa is highly skewed, with five coastal nations (Angola, Kenya, Nigeria, South Africa, and Sudan) accounting for more than 70 percent of the continent's GDP (Figure 1.4). Poor infrastructure prevents integrated and more productive economic networks from emerging. Integrating the leading regions with those that lag is crucial to ensuring that growth spills over and creates opportunities and shared prosperity across the continent. Regional integration brings both supply-side and demand-side benefits. On the supply side it facilitates the movement of factors of production to create the capacity and scale lacking in small countries. On the demand side it provides access to larger markets. Opportunities for cross-border trade in food products, basic manufactures, and services remain untapped, and the benefits could be enormous.[2] Integration is necessary at all geographic scales: local integration increases earning opportunities for farmers and small firms in lagging parts of a country, regional integration scales up supply capacity among countries with individually insufficient physical and human capital, and global integration creates access to demand in wealthier world regions.

Figure 1.4 Spatial Distribution of GDP

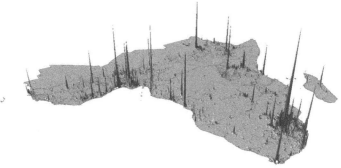

Source: Calculated from Ghosh et al. 2010.

Table 1.1 Density of Paved Roads in Sub-Saharan Africa, Compared with Other Low-Income Countries

	Sub-Saharan Africa low-income countries	Other low-income countries
Density by area (km/1,000 km²)	10.7	37.3
Density by population (km/1,000 population)	269.1	700.7
Density by GDP per capita (km/US$ billion)	663.1	1,210.0

Source: Carruthers, Krishnamani, and Murray 2008.

Removing nonphysical trade barriers will be important to promoting integration. However, as the African Infrastructure Country Diagnostic (Foster and Briceño-Garmendia 2010) has shown, and this study will elaborate, integration cannot occur without far better physical infrastructure, especially roads. Africa's road infrastructure is deficient by almost any measure, even when compared with countries at similar income levels in other regions. Road density, the appropriateness of which depends on both population density and economic activity, is an oft-cited but imperfect proxy for connectivity. The road network in Sub-Saharan Africa is significantly smaller than that of any other region, with only 204 kilometers of road per 1,000 square kilometers of land area, nearly one-fifth the world average, and less than 30 percent of the next worst region, South Asia. Only 25 percent of Sub-Saharan Africa's roads are paved, compared with about 50 percent in South Asia.[3] When comparing only Sub-Saharan Africa's low-income countries with low-income countries of the rest of the world, similar patterns arise (Table 1.1).

Requirements for Improving Connectivity

The need to enhance connectivity in Africa is broadly recognized. As a consequence governments and donors in Sub-Saharan Africa have devoted considerable resources to the construction and rehabilitation of roads. An emphasis on transport infrastructure is also evident in the lending patterns of the World Bank, which commits a larger share of resources to transport infrastructure than to education, health, and social services combined (World Bank 2007). In fiscal year 2013, the World Bank's total transport commitments amounted to US$5.9 billion, and rural and interurban roads remained the largest subsector, accounting for 60 percent of total lending (World Bank 2014b).

Proposals for ambitious future transport linkages abound. The Africa Infrastructure Country Diagnostic identified spending needs on road infrastructure amounting to more than US$18 billion dollars per year. NEPAD has constructed a large proposal for road construction and improvement of nine trans-African highways spread over the continent. These nine highways include nearly 60,000 kilometers of road; about 25 percent of the network consists of "missing links" that need to be built from scratch or significantly improved, at an estimated cost of US$4.2 billion (AfDB and UNECA 2003). Given limited resources, the full funding of such grandiose plans is doubtful. Furthermore, there is little knowledge or consensus about exactly where investments might be transformative in creating new opportunities for regional trade. Plans are often aspirational and based on assumptions of perceived significance rather than on empirical quantitative assessments of potential productivity and relative efficiency.

The policy implication is straightforward: prioritization is crucial because the gap between requirements and available funds is vast. A review by Dercon and Lee (2012, 4) states the problem concisely in its conclusion that there is still "limited understanding of . . . what needs to be done to avoid investing massive resources in infrastructure which does not result in economic growth."

Prioritization will require reliable estimates of the benefits and costs of specific proposed investments. Estimating construction costs is not necessarily simple, but it is primarily an engineering issue. Evaluating benefits, however, is analytically considerably more challenging. One approach is simply to forecast the traffic increase or time savings that would come from building or upgrading a road. But, apart from the difficulty of accurately forecasting future developments, this method fails to capture the potential indirect benefits of increasing a region's connectivity, including economic multiplier effects and better access for the affected populations to social infrastructure. An alternative approach in the literature, and the one adopted in this study, uses actual data from existing roads to estimate the roads' benefits, and then uses these estimates to forecast benefits from proposed construction or upgrading.

This approach also faces challenges, including the difficulty of obtaining data that accurately reflect the conditions of the roads and the cost of traveling along them. These data gaps are always a concern when dealing with road infrastructure—the quality of which is constantly in flux—but they are especially a challenge in Africa where infrastructure assessments are infrequent and rural roads are often unaccounted for. The second econometric challenge, as discussed later in this chapter in the summary of chapter 2, is how to overcome several potential sources of endogeneity, which, if not accounted for, will bias estimates of benefits.

Prioritization also requires evaluating the potential costs of building roads over and above the cost of physical construction. The most significant of these costs is probably the impact on forests. Deforestation in the Amazon is clearly associated with road construction (Moran 1993). Recent World Bank assessments suggest that major threats to the Congo region are likely to occur not from forest harvesting, but rather from induced (and often unintended) land-use changes deriving from activities such as agriculture and mining facilitated by transport investments (Megevand 2013). Given its economic significance, a careful assessment and evaluation of impacts and trade-offs is necessary to prevent excessive damage to this global asset. This study applies its tools to this question as well.

Literature on the Relationship between Infrastructure and Welfare

Despite the centrality of transport costs in economic theory (especially trade economics, spatial economics, and the new economic geography), establishing causal links between infrastructure investment and economic growth has proved to be elusive. The debate on this topic has been fostered by limited evidence of a causal relationship and conflicting evidence provided by different studies (Gunasekara, Anderson, and Lakshmanan 2008). The early empirical literature was inspired by the seminal work of Aschauer (1989) and sought to estimate the effects of public investment (especially in highways) on overall employment or growth using time series data. These estimates are deemed to be implausibly large because of the nonstationary nature of the data and the possible endogeneity of location of public investments. Subsequent work has sought to correct these problems, with generally mixed results. One approach relates regional-level public expenditure data to changes in agricultural productivity (for example, Fan, Hazell, and Thorat 2000). However, this approach does not disaggregate the components of public spending and so cannot inform investment decisions. Another approach, developed by Straub and Terada-Hagiwara (2011), examines infrastructure in a growth-accounting framework and finds that increases in infrastructure were correlated significantly with growth. But the evidence demonstrated that this was mostly the result of general increases in capital and skills, and evidence of infrastructure's direct impact on productivity was again inconclusive. A correlation between infrastructure and output is not sufficient to show its contribution to growth.

Researchers have also examined the effects of road infrastructure and transport capital investments on aggregate productivity (usually measured by GDP or personal income), output elasticity, and productivity in developed countries (Aschauer 1989; Lakshmanan and Anderson 2002, 2007; Chandra and Thompson 2000; Demetriades and Mamuneas 2000; Annala and Perez 2001; Foster and Araujo 2004; Ihori and Kondo 2001; Lokshin and Yemtsov 2003; Nadiri and Mamuneas 1996; Shirley and Winston 2004), and in developing countries (Storeygard 2014; Morrison-Paul et al. 2004; Lokshin and Yemtsov 2003; Feltenstein and Ha 1995). The results, however, remain ambiguous with conflicting evidence of impacts in both developed and developing countries. To a large extent, the contradictory evidence and the ensuing debates are a consequence of endogeneity, identification, and reverse causality problems.

A set of papers has used various identification strategies to shed light on the impact of large transport infrastructure improvements (Michaels 2008; Donaldson, forthcoming; Datta 2012; Faber 2014; Banerjee, Duflo, and Qian 2012). One approach is to use panel data estimation methods (Dercon et al. 2008; Khandker and Koolwal 2011). Regrettably, however, panel data on transport costs with adequate observations over time are rare, especially for developing countries, and not available for countries used in this study, specifically Nigeria, the Democratic Republic of Congo, and other West and Central African countries. Another approach is to use spatial panel data with natural experiments that exploit the historical context of transport infrastructure (Jedwab and Moradi 2012; Banerjee, Duflo, and Qian 2012). Others have used difference-in-difference (Datta 2012) or difference-in-difference with an instrumental variable (Faber 2014). In the absence of natural experiments and panel data, numerous studies have attempted to capture exogenous variations in transport costs by incorporating exogenous geographic features (Jacoby and Minten 2009; Shrestha 2012; Emran and Hou 2013).

A number of studies have shown the benefits of greater connectivity, for instance, proximity to rural roads (Dercon et al. 2008; Bosker and Garretsen 2012). Other papers look at the mechanisms through which transport costs impact well-being. One impact is that reducing transport costs leads to greater access to markets (easier and cheaper access to both purchased inputs and markets for outputs) as well as to decreases in both trade costs and interregional price gaps (Donaldson, forthcoming; Casaburi, Glennerster, and Suri 2013). Better access further affects the input and output prices of crops (Khandker, Bakht, and Koolwal 2006; Minten and Kyle 1999; Chamberlin et al. 2007;

Stifel and Minten 2008) as well as land value (Jacoby 2000; Shrestha 2012; Donaldson, forthcoming; Gonzalez-Navarro and Quintana-Domeque 2010). Not surprisingly, the literature also finds that access to good-quality roads facilitates economic diversification (Gachassin, Najman, and Raballand 2010; Fan, Hazell, and Thorat 2000; Mu and van de Walle 2007). Another improvement to well-being is that reduced transport costs can insure farmers against negative shocks (Burgess and Donaldson 2012).

The Structure of the Report

This report advances the current state of knowledge by modeling the economic and behavioral effects of reducing transport costs through road construction or upgrading, and then applying some unique data sets and novel methodologies to quantify the impacts. These methodologies and data sets are designed to overcome some of the problems that the studies mentioned previously have encountered. By tackling the problem of benefit and cost estimation in this way while also taking important steps to ensure the results reflect a causal impact and not a simple correlation, this study is intended to inform the continuing debate (among planners, policy makers, and academics) about the role of transport investments in economic growth.

Following this introductory chapter, the report comprises five additional chapters. Chapter 2 describes in detail the general methodological and data-related challenges, and how they are tackled, then applies this methodology to the estimation of benefits, and finally demonstrates how this approach can be used to evaluate an actual proposal for road construction and upgrading. Chapters 3 through 5 use different behavioral models but a similar econometric methodology to investigate several interrelated topics that are especially relevant for Africa: (1) how a reduction in transport costs may produce benefits for farmers by inducing the uptake of modern production technologies, which as noted previously, is key to reversing the growing productivity gap between Africa and other regions; (2) the relationship between conflict and the benefits derived from reducing transport costs; and (3) quantification of the negative environmental externalities from road construction. The last chapter summarizes the main messages, caveats, and the road forward. Finally, the reader will find a Geospatial appendix with detailed information of the GIS techniques employed. Each chapter is introduced in more detail below.

Chapter 2. Welfare Effects of Road Infrastructure
This chapter describes the methodological approach used throughout the study and demonstrates how it can be used to assess the differential development impacts of alternative road construction and to prioritize various proposals.

This methodology tackles the key challenges of estimating the impact of road networks on economic activity, which are (1) obtaining quality data that accurately reflect the condition of the roads and the costs of traveling along them; and (2) overcoming the potential sources of endogeneity[4] arising from the nonrandom placement of roads, spatial sorting of households, and the geographic emergence of markets.

This study adds significant value to the literature in that it aims to overcome the data challenge by estimating as accurately as possible the actual cost of transporting goods to market, using geographic information system software, a carefully constructed road network, and a unique algorithm for estimating the costs of moving along this road network. Taking into account road quality, type of paving, roughness of the terrain, and local transport input costs (for instance, price of fuel, price of a truck, wages), the measure of transport cost to market used here is arguably the best possible given existing technology.

The potential endogeneity bias is handled using a novel instrumental variable (IV), termed the "natural path." The natural path estimates a route that gives the shortest time it would take to walk between any two points on a map, given the topography of the land and in the absence of any transportation infrastructure. This instrument is an improvement over the straight-line (Euclidean distance) instruments frequently used in the literature because it more accurately captures what straight-line instruments attempt to estimate—that is, the most logical route connecting two points—without taking into account other, bias-causing economic benefits. Although natural path and Euclidean IVs are highly correlated in a featureless terrain, the difference is significant where the land has topographic characteristics that affect mobility.

Furthermore, recognizing that no perfect measure of economic well-being exists, a variety of outcome metrics are used, including crop revenue, livestock sales, nonagricultural income, probability of nonagricultural employment, probability of being multidimensionally poor, wealth, and local GDP for Nigeria. The household welfare indicators come from data generated by two particularly rich surveys: the Living Standards Measurement Study–Integrated Survey of Agriculture and the Demographic and Health Survey. The combination of these data with the methodology described above gives a more complete picture of the extent to which household welfare is expected to improve with a given reduction in transport costs.

This approach generates elasticities that enable a forecast to be made of the economic impact of the construction of future roads or the improvement of any portion of the current network. With these more accurate estimates of the expected impact on economic activity and the welfare of the rural poor, an evaluation of the relative benefits of alternative road construction proposals becomes possible. This knowledge will enable decision makers to prioritize construction of those roads that would have the biggest impact on spurring economic growth and reducing poverty in the region. This technique is illustrated through an evaluation of one of the regional projects in NEPAD's proposal. It should be understood, however, that this approach does not allow comparison of road investments with alternative uses of public funds, nor are multimodal transport investments modeled. An additional caveat is that realization of the benefits of road construction depends crucially on the presence of enabling factors. One of these factors, emphasized in chapter 4, is peace and security, but a full list would include all conditions contributing to a good business environment.

Chapter 3. Impact of Transport Cost on Technology Adoption

This chapter addresses one of the most vexing questions facing efforts to trigger transformational change in African agriculture: how to encourage greater uptake of modern productivity-enhancing technologies (fertilizers, seeds, irrigation, and mechanization) by farmers. This is a fundamental challenge, and one that has yet to be effectively met, despite decades of attempts by development agencies, nongovernmental organizations, and African governments. Many diagnoses have been made of the underlying cause of the failure of farmers to adopt these technologies, including risk aversion, insufficient incentives to invest when there are high taxes on production, high distribution margins in small oligopolistic input markets, lack of rural financial markets, regulatory barriers, risks involved in making capital investments, and farmers' ignorance (due partially to ineffective extension). It is likely that each of these may play some role. Chapter 3 suggests another significant contributory factor, not mutually exclusive with some or all of the other impediments mentioned.

To frame the analysis, this chapter outlines a minimalist model of technology adoption with transport costs. It is assumed that there are fixed costs, or minimum threshold costs, to adopting more modern agricultural inputs (for instance, a minimum rental or purchase price of tractors, harvesters, and planters; learning costs for new production techniques; and so forth). The fixed costs create a hurdle that households must overcome to adopt the more productive technology. Transport costs influence the returns to technology adoption and thereby create heterogeneity of responses, that is, farmers choose different levels of technology based on their transport costs. Constraints on technology adoption can impede entry into markets and lock farmers into traditional, low-input modes of production, while variations in transport costs generate differences in returns. This model generates three testable hypotheses: (1) adoption of new technologies will be more pervasive if transport costs are lower; (2) reductions in transport costs will have a larger impact on the marketed output of farmers who are using modern farming techniques; and (3) farmers using more modern technologies are likely to be better integrated into markets. The study tests these hypotheses using methodologies and data similar to those in chapter 2.

Chapter 4. Role of Transport Infrastructure in Conflict-Prone and Fragile Environments: Evidence from the Democratic Republic of Congo

The rehabilitation of damaged road infrastructure is an overarching investment priority among donors and governments in conflict-prone and fragile states. All of NEPAD's proposed new highways (AfDB and UNECA 2003) pass through fragile states (as defined by the Organisation for Economic Co-operation and Development), and in most cases, the portions of the roads most in need of rehabilitation lie within these countries. However, little empirical evidence is available on the direct causal impact of access to markets on well-being in fragile situations when the risks of reversion to conflict are high, and even less evidence covers the combined impact of transport costs and conflict. This chapter addresses the question of how the presence of conflict may affect the degree to which reduction in transport costs produces benefits. A two-stage model of conflict is constructed, in which agents choose to be either farmers or rebels. Their choice is influenced by their economic "distance" from the market, which is a function of transport costs. Rebels must choose where to focus their attacks—on the market area or on farm households—with implications for the behavior and welfare of farmers.

The chapter empirically tests the predictions from this model that (1) an increase in transport costs diverts attacks from goods sold at the market to subsistence goods (at the farmer's location); and (2) in certain situations, lower transport costs would induce a switch from farming to rebellion and vice versa. Econometric testing of these predictions is hampered not only by the data and endogeneity issues described already, but also by two additional problems: (1) conflict itself is endogenous (is welfare low because there is conflict nearby, or is conflict generated by poverty?) and (2) depending on other conditions, lower transport costs could either induce a switch from farming to rebellion or vice versa. To resolve the first issue, the estimation technique uses ethnic fractionalization as an IV. For the second, a novel technique is used to measure conflict by calculating a kernel that takes into account the intensity of conflict at a given location, as well as that in nearby areas, with weights decreasing with distance.

Chapter 5. Road Improvements and Deforestation in the Congo Basin Countries

Chapter 5 estimates the effects of road access on deforestation and associated change in biodiversity in eight Congo Basin countries (Burundi, Cameroon, Central African Republic, the Democratic Republic of Congo, the Republic of Congo, Equatorial Guinea, Gabon, and Rwanda), then uses the estimates to simulate the effect of improving road segments on deforestation in northeast Democratic Republic of Congo. Saving the forests in the Congo Basin from the fate of the Asian and Amazon rain forests is important for the region itself because the forests provide hugely important eco-services. They are critical in hydrological regulation and are the source of much of the rainfall in the Sahel and East Africa (Megevand 2013). They also host an array of charismatic and threatened species, such as the lowland gorilla, chimpanzee, and the African forest elephant. Protecting these rain forests is also significant for the fight against global climate change since they contain 30–40 gigatons of carbon—8 percent of the world's forest carbon, the equivalent of three to five years of world emissions of carbon dioxide equivalent. Deforestation is clearly linked with opening roads through the forest. Given the importance of roads for raising the welfare of the populations of the region, which are some of the poorest in the world, avoiding road building altogether is not desirable. But it is clearly in everyone's best interest to place roads so that they provide maximum benefits while causing minimum collateral damage to the ecosystem. Therefore, this chapter addresses critical issues for decision makers: First, what factors and conditions influence the degree to which building a road will result in deforestation in the Congo region? Second, how can the likely damage from a given road be quantified in a way that will allow it to be compared with other options for placement or improvement of the road, or of an alternative road?

Similar to the preceding two chapters, chapter 5 first develops a theoretical model that considers the potentially adverse impact of traditional road improvement planning, in which decision making is sequential: decisions on road improvement projects in an area are made first, followed by environmental impact assessments that seek to mitigate the impact of forest clearing by strengthening environmental management rather than affecting the selection of projects. The modeling exercise shows why coordinated infrastructure planning in such a sequential decision regime, while otherwise desirable, may actually reduce welfare because it increases deforestation and the associated ecological impacts. The chapter then investigates the impact of road building on deforestation using a recently published, first of its kind data set of high-resolution, consistently derived estimates of global forest clearing. It also constructs an index of ecosystem risks and overlays the basin-wide road network on this map to provide a first-order guide to risk assessment for proposed road corridor improvements. The index is arguably unique and combines different measures and criteria for determining risk in a single index. Finally, it combines the ecological risk indicator with pixel-level predictions of forest clearing produced by road upgrading. The result is a high-resolution map of expected risks for road upgrading in road segments, corridors, and regional networks. The chapter illustrates the implications in a detailed assessment for the northeastern portion of the Democratic Republic of Congo. It combines the estimates of road construction benefits with the environmental impacts. The approach developed here thus provides a novel and pioneering approach to integrating environmental concerns with economic priorities.

New Techniques for Answering Questions in Infrastructure Economics

The chapters address different aspects of the impact of transport on welfare and are not intended to comprise a coherent story line. Taken together, however, this report is a comprehensive study of the local effects of improving a country's local or national road network. For illustrative purposes, two countries are chosen as case studies to apply the techniques developed here—the Democratic Republic of Congo and Nigeria. However, the methodology could be extended to any developing country. Although the results in the subsequent chapters are meant to be useful for policy makers within their respective countries, it is hoped that the full contribution of this report is not limited to the handful of results and simulations that space considerations allow for, but that the methodologies developed and applied here can inform future policy decisions more broadly.

Notes

1. For example, Mason and Jayne (2009) find that for urban consumers in Zambia, average expenditures on food were 45–55 percent of their incomes.
2. World Bank (2012) demonstrates that it is easier for Africa to trade with the rest of the world than with itself.
3. Though the discussion is about road networks this report later derives travel costs, which is a more accurate measure of connectivity.
4. See Emran and Hou (2013), which discusses these three sources of bias in the context of rural China. Regarding nonrandom placement, roads tend to be placed to connect areas of high economic activity; therefore, even if the presence of a road and welfare are positively correlated, the direction of causality is not clear. Regarding spatial sorting, the location of families near roads may be nonrandom as well; for example, if high-potential households are more likely to settle near roads, then the correlation between road presence and high incomes may not indicate that the road actually caused the affluence. Regarding market emergence, towns and market centers are likely to arise in areas of high economic potential, and roads may follow, again producing a correlation that does not indicate causation.

References

AfDB and UNECA (African Development Bank and United Nations Economic Commission for Africa). 2003. "Review of the Implementation Status of the Trans African Highways and the Missing Links, Volume 1: Main Report." SWECO International AB, and Nordic Consulting Group AB, Sweden.

Annala, C., and P. Perez. 2001. "Convergence of Public Capital: Investment among the United States, 1977–1996." *Public Finance and Management* 1 (2): 214–29.

Aschauer, David Alan. 1989. "Is Public Expenditure Productive?" *Journal of Monetary Economics* 23 (2): 177–200.

Banerjee, Abhijit, Ester Duflo, and Nancy Qian. 2012. "On the Road: Access to Transportation Infrastructure and Economic Growth in China." NBER Working Paper 17897, National Bureau of Economic Research, Cambridge, Massachusetts.

Bosker, Maarten, and Harry Garretsen. 2012. "Economic Geography and Economic Development in Sub-Saharan Africa." *World Bank Economic Review* 26 (3): 443–85.

Burgess, Robin, and Dave Donaldson. 2012. "Railroads and the Demise of Famine in Colonial India." Working Paper, London School of Economics, MIT, and National Bureau of Economic Research.

Byerlee, Derek, Andres F. Garcia, Asa Giertz, and Vincent Palmade. 2013. *Growing Africa: Unlocking the Potential of Agribusiness*. Washington, DC: World Bank. http://documents.worldbank.org/curated/en/2013/03/17427481/growing-africa-unlocking-potential-agribusiness-vol-1-2-main-report.

Carruthers, Robin, Ranga R. Krishnamani, and Siobhan Murray. 2008. "Improving Connectivity: Investing in Transport Infrastructure in Sub-Saharan Africa." Background Paper 7 for Africa Infrastructure Country Diagnostic, World Bank, Washington, DC.

Casaburi, Lorenzo, Rachel Glennerster, and Tavneet Suri. 2013. "Rural Roads and Intermediated Trade: Regression Discontinuity Evidence from Sierra Leone." SSRN. http://ssrn.com/abstract=2161643 or http://dx.doi.org/10.2139/ssrn.2161643.

Chamberlin, Jordan, Muluget Tadesse, Todd Benson, and Samia Zakaria. 2007. "An Atlas of the Ethiopian Rural Economy: Expanding the Range of Available Information for Development Planning." *Information Development* 23 (2-3): 181–92.

Chandra, Amitabh, and Eric Thompson. 2000. "Does Public Infrastructure Affect Economic Activity? Evidence from the Rural Interstate Highway System." *Regional Science and Urban Economics* 30 (4): 457–90.

Datta, Saugato. 2012. "The Impact of Improved Highways on Indian Firms." *Journal of Development Economics* 99 (1): 46–57.

Deininger, Klaus, and Derek Byerlee. 2011. *Rising Global Interest in Farmland: Can It Yield Sustainable and Equitable Benefits?* Washington, DC: World Bank.

Demetriades, Panicos, and Theofanis Mamuneas. 2000. "Inter-Temporal Output and Employment Effects of Public Infrastructure Capital: Evidence from 12 OECD Economies." *Economic Journal* 110 (465): 687–712.

Dercon, Stefan, Daniel O. Gilligan, John Hoddinott, and Tassew Woldehanna. 2008. "The Impact of Agricultural Extension and Roads on Poverty and Consumption Growth in Fifteen Ethiopian Villages." IFPRI Discussion Paper 00840, International Food Policy Research Institute, Washington, DC.

Dercon, Stefan, and Stevan Lee. 2012. "Evidence on Infrastructure and Growth." *Growth Research News*, December. https://www.gov.uk/government/uploads/system/uploads/attachment_data/file/197855/growth-research-news-1212.pdf.

Donaldson, David. Forthcoming. "Railroads of the Raj: Estimating the Impact of Transportation Infrastructure." *American Economic Review*.

Emran, Shahe, and Zhaoyang Hou. 2013. "Access to Markets and Household Consumption: Evidence from Rural China." *Review of Economics and Statistics* 95 (2): 682–97.

Faber, Benjamin. 2014. "Trade Integration, Market Size, and Industrialization: Evidence from China's National Trunk Highway System." *Review of Economic Studies.* doi:10.1093/restud/rdu010.

Fan, S., P. Hazell, and S. Thorat. 2000. "Government Spending, Growth and Poverty in Rural India." *American Journal of Agricultural Economics* 82 (4): 1038–51.

Feltenstein, Andrew, and Jiming Ha. 1995. "The Role of Infrastructure in Mexican Economic Reform." *World Bank Economic Review* 9 (2): 287–304.

Foster, Vivien, and Maria Caridad Araujo. 2004. *Does Infrastructure Reform Work for the Poor? A Case Study from Guatemala.* Washington, DC: World Bank.

Foster, Vivien, and Cecilia Briceño-Garmendia. 2010. *Africa's Infrastructure: A Time for Transformation.* Washington, DC: World Bank.

Gachassin, Marie, Boris Najman, and Gaël Raballand. 2010. "The Impact of Roads on Poverty Reduction: A Case Study of Cameroon." Policy Research Working Paper 5209, World Bank, Washington, DC.

Ghosh, T., R. L. Powell, C. D. Elvidge, K. E. Baugh, P. C. Sutton, and S. Anderson. 2010. "Shedding Light on the Global Distribution of Economic Activity." *Open Geography Journal* 3 (1): 148–61.

Gonzalez-Navarro, Marco, and Climent Quintana-Domeque. 2010. "Roads to Development: Experimental Evidence from Urban Road Pavement." UC Berkeley and Universitat d'Alacant. http://dx.doi.org/10.2139/ssrn.1558631.

Gunasekara, Kumudu, William P. Anderson, and T. R. Lakshmanan. 2008. "Highway Induced Development: Evidence from Sri Lanka." *World Development* 36 (11): 2371–89.

Haggblade, Steven. 2013. "Unscrambling Africa: Regional Requirements for Achieving Food Security." *Development Policy Review* 31 (2): 149–76.

Ihori, Toshihiro, and Hiroki Kondo. 2001. "The Efficiency of Disaggregate Public Capital Provision in Japan." *Public Finance and Management* 1 (2): 161–82.

Ivanic, M., and W. Martin. 2014. "Sectoral Productivity Growth and Poverty Reduction: National and Global Impacts." Unpublished, Development Economics Group, World Bank, Washington, DC.

Jacoby, Hanan G. 2000. "Access to Markets and the Benefits of Rural Roads." *Economic Journal* 110 (465): 713–37.

Jacoby, Hanan G., and Bart Minten. 2009. "On Measuring the Benefits of Lower Transport Costs." *Journal of Development Economics* 89 (1): 28–38.

Jedwab, Remi, and Alexander Moradi. 2012. "Colonial Investments and Long-Term Development in Africa: Evidence from Ghanaian Railroads." Unpublished, George Washington University, Washington, DC; STICERD, London School of Economics; and University of Sussex.

Khandker, Shahidur R., Zaid Bakht, and Gayatri Koolwal. 2006. "The Poverty Impact of Rural Roads: Evidence from Bangladesh." Policy Research Working Paper 3875, World Bank, Washington, DC.

Khandker, Shahidur R., and Gayatri B. Koolwal. 2011. "Estimating the Long-Term Impacts of Rural Roads: A Dynamic Panel Approach." Policy Research Working Paper 5867, World Bank, Washington, DC.

Lakshmanan, T. R., and William P. Anderson. 2002. "Transportation Infrastructure, Freight Services Sector, and Economic Growth." A White Paper Prepared for the U.S. Department of Transportation, Federal Highway Administration, Center for Transportation Studies, Boston University.

———. 2007. "Transport's Role in Regional Integration Processes." In *Market Access, Trade in Transport Services and Trade Facilitation*, ECMT Round Tables, No. 134, 45–72. Paris: OECD Publishing.

Lokshin, Michael, and Rusian Yemtsov. 2003. "Evaluating the Impact of Infrastructure Rehabilitation Projects on Household Welfare in Rural Georgia." Policy Research Working Paper 3155, World Bank, Washington, DC.

Mason, Nicole M., and Thomas S. Jayne. 2009. "Staple Food Consumption Patterns in Urban Zambia: Results from the 2007/2008 Urban Consumption Survey." Food Security Collaborative Policy Brief 56810, Department of Agricultural, Food, and Resource Economics, Michigan State University, East Lansing, Michigan.

Megevand, Carole, with Aline Mosnier, Joël Hourticq, Klas Sanders, Nina Doetinchem, and Charlotte Streck. 2013. *Deforestation Trends in the Congo Basin: Reconciling Economic Growth and Forest Protection*. Washington, DC: World Bank.

Michaels, Guy. 2008. "The Effect of Trade on the Demand for Skill: Evidence from the Interstate Highway System." *Review of Economics and Statistics* 90 (4): 683–701.

Minten, Bart, and Steven Kyle. 1999. "The Effect of Distance and Road Quality on Food Collection, Marketing Margins, and Traders' Wages: Evidence from the Former Zaire." *Journal of Development Economics* 60 (2): 467–95.

Moran, Emilio F. 1993. "Deforestation and Land Use in the Brazilian Amazon." *Human Ecology* 21 (1): 1–21.

Morrison-Paul, C., R. Nehring, D. Banker, and A. Somwaru. 2004. "Scale Economies and Efficiency in U.S. Agriculture: Are Traditional Farms History?" *Journal of Productivity Analysis* 22 (November): 185–205.

Mu, Ren, and Dominique van de Walle. 2007. "Rural Roads and Poor Area Development in Vietnam." Policy Research Working Paper 4340, World Bank, Washington, DC.

Nadiri, M. Ishaq, and Theofanis P. Mamuneas. 1996. "Contribution of Highway Capital to Industry and National Productivity Growth." Report for Apogee Research Inc. and the Federal Highway Administration Office of Policy Development.

Nash, John, Naomi Halewood, and Samia Melhem. 2013. *Unlocking Africa's Agricultural Potential: An Action Agenda for Transformation*. Africa Region Sustainable Development Series. Washington, DC: World Bank. http://documents.worldbank.org /curated/en/2013/04/17630278/unlocking-africas-agricultural-potential-action-agenda -transformation.

Pardey, Philip G., Julian M. Alston, and Roley R. Piggot. 2006. "Agricultural R&D in the Developing World: Too Little, Too Late?" International Food Policy Research Institute, Washington, DC.

Shirley, Chad, and Clifford Winston. 2004. "Firm Inventory Behavior and the Returns from Highway Infrastructure Investments." *Journal of Urban Economics* 55 (2): 398–415.

Shrestha, Slesh A. 2012. "Access to the North-South Roads and Farm Profits in Rural Nepal." University of Michigan.

Stifel, David, and Bart Minten. 2008. "Isolation and Agricultural Productivity." *Agricultural Economics* 39 (1): 1–15.

Storeygard, Adam. 2014. "Farther on down the Road: Transport Costs, Trade and Urban Growth in Sub-Saharan Africa." Policy Research Working Paper 6444, World Bank, Washington, DC.

Straub, Stéphane, and Akiko Terada-Hagiwara. 2011. "Infrastructure and Growth in Developing Asia." *Asian Development Review* 28 (1): 119–56.

World Bank. 2007. *Evaluation of World Bank Support to Transportation Infrastructure.* Internal Evaluation Group. Washington, DC: World Bank.

———. 2012. *Africa Can Help Feed Africa: Removing Barriers to Regional Trade in Food Staples.* Washington, DC: World Bank.

———. 2014a. "Africa's Pulse: Decades of Sustained Growth is Transforming Africa's Economies." World Bank, Washington, DC. http://www.worldbank.org/en/region /afr/publication/africas-pulse-decades-of-sustained-growth-is-transforming-africas -economies.

———. 2014b. "Transport: Sector Results Profile—Sustainable Transport for All: Helping People to Help Themselves." World Bank, Washington, DC.

Chapter 2

Welfare Effects of Road Infrastructure

Introduction

Transport infrastructure is deemed to be central to development and consumes a large fraction of development assistance. Yet debate continues about the economic impact of road projects. Much of this debate ensues from the challenges inherent in accurately estimating the impact of roads on welfare. Given the nonrandom locations of households, markets, and the roads connecting them, disentangling whether the road brought economic growth to the area, or the economic potential of the area led to the road being built there in the first place, becomes challenging. This chapter addresses these and other issues using the best data and techniques available. It also proposes an approach to assessing the different development impacts of alternative projects and prioritizing various proposals. Recognizing that there is no perfect measure of economic well-being, a variety of outcome metrics are used, including crop revenue, livestock sales, nonagricultural income, probability of nonagricultural employment, the probability of being multidimensionally poor, wealth, and local GDP. The focus of the subsequent analysis is on Nigeria, but the approach could be extended to other parts of the world. The chapter is divided into two major sections, with the first describing the challenges of estimating benefits of road construction and upgrading, then illustrating how these challenges can be addressed. The second section uses the results from the first section (the estimated relationship between transport cost and welfare) to illustrate how they can be applied to evaluating and prioritizing realistic proposals.

Governments and donors in Sub-Saharan Africa have devoted considerable resources to the construction and rehabilitation of roads. It is widely accepted that roads, although expensive, facilitate the creation of and the participation in markets and are therefore central to development. Benefits from new transportation infrastructure can come in many forms: economic benefits, including the reduction of business costs and enhanced access to markets; social benefits, including improved social cohesion, faster diffusion of information, and better access to schools and hospitals; and political benefits, such as the pleasing of certain constituencies, among others.

Africa has the lowest density of roads in the world, with 204 kilometers of road per 1,000 square kilometers of land area, nearly one-fifth the world average and less than 30 percent of the next worst region, South Asia. Starting from such a low base, the potential for growth due to improvements in transportation infrastructure is presumed to be especially large in Africa. In the context of African agriculture, productivity growth lags far behind other regions in large part because of low levels of use of modern inputs (World Bank 2013), as well as limited connectivity to markets.

Challenges in Estimating the Economic Effects of Transport Cost Reductions

Notwithstanding the general acceptance of the efficacy of roads as a develop-ment tool, the actual empirical results from the body of research linking roads—and reduced transport costs—to economic well-being remain surprisingly ambiguous. To a large extent, the contradictory evidence and ensuing debates are a consequence of the challenges inherent in accurately estimating the impact of roads on welfare. As noted in chapter 1, two key challenges in estimating the impact of road networks on economic activity contribute to this ambiguity. The first is the difficulty of obtaining data that accurately reflect the conditions of the roads and the challenges of traveling along them. This is a particular concern in Africa, where infrastructure assessments are infrequent and rural roads are often unaccounted for.

The second challenge stems from the nonrandom locations of households, markets, and the roads connecting them. For example, wealthier households might locate closer to markets on the basis of household characteristics that are unobservable to the researcher. In addition, roads tend to be built to connect major economic activities, for example, linking cities, markets, mines, or areas of high agricultural productivity, or they may be built in areas that lack develop-ment in the expectation of an economic stimulus. So, the fact that roads are near wealthier regions does not mean that the roads caused those areas to be richer, and failure to take this endogeneity into account in estimating the benefits of roads will lead to statistically biased results. As discussed below, this endogene-ity is addressed with the use of an instrumental variable (IV).

How This Study Differs from Others on This Topic
One major objective of this research is to overcome the challenges outlined above and to estimate more accurately the real value of infrastructure that improves connectivity in Africa. In doing so, it also develops a simple simula-tion approach that can be used to evaluate the benefits of specific investments. This method is not a substitute for a full benefit-cost analysis that includes

all externalities, but it does have a very important advantage. Traditional assessments of road impacts typically focus only on the benefits to users of the road and ignore the economywide impacts that would result from the induced economic activity and trade. The approach developed here captures not only benefits to direct, local users of the road, but also benefits that accrue to the wider economy in the form of multiplier effects from changes in transport costs. It can thus be viewed as a complement to the more traditional ways of measuring benefits from transport infrastructure.

This research works to resolve the data challenge mentioned above by carefully georeferencing the existing Nigerian road network and estimating costs of moving along the network. To construct a measure of travel costs from producers and households to markets in Nigeria, geographic information system road network data[1] are combined with road survey data from the Nigeria Federal Roads Maintenance Agency (FERMA) and the World Bank's Fadama project.[2] The resulting map of Nigeria's road system is shown in map 2.1. The Highway Development Management Model (HDM-4) programming tool is then used to estimate a cost per ton per kilometer of traversing this road, taking into account the roughness of the terrain, quality and condition of the road, and country-level factors (such as the price of fuel, average quality of the fleet, the price of a used truck, and wages). Finally, the total travel cost to the cheapest market via the optimal route from each location is calculated using an iterative

Map 2.1 Nigeria's Road Network, by Condition

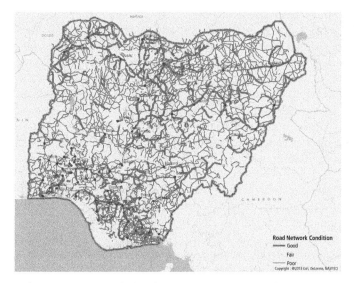

Sources: Data from DeLorme; Nigeria Federal Roads Maintenance Agency; and World Bank 2013.

cost-minimizing process in which every possible travel path to every available market is calculated, and the least-cost one was chosen. Arguably the resulting measure of transport cost is the most accurate possible given existing information. This study then develops an accurate estimate of the degree to which a road investment will reduce transport costs, which is critical in quantifying the welfare benefits of potential projects.

The approach adopted in this research to correct for the endogeneity of roads is to estimate the impact using IVs calculated for this purpose.[3] The use of IVs is a methodology common in economic research to investigate whether variable A "causes" variable B, when there is a strong possibility that A is, in fact, influenced by B through some reverse channel of causation. In such cases, if one can find a variable C that is well correlated with A but not influenced by B (or vice versa),[4] then C can be used as an "instrument" for A in regressions. For this study, the IV must be correlated with transport costs, and must only affect the outcome variable (various measures of welfare) through its effect on transport costs, not directly. The variable that has most often been used to instrument for transport cost in researching the impacts of roads on welfare is the straight-line or Euclidean distance from the household to the market. This measure is arguably uncorrelated with a road's benefits[5] but intuitively must be at least somewhat correlated with travel time.

However, while useful, the straight-line measure ignores a major determinant of travel time and cost—the topography of the land. Even if two points are close together, they may not be easily accessible (for example, if separated by a canyon or steep terrain), making it expensive to travel from one point to the other, even if a road approximating that straight line existed. To improve upon Euclidean distance, an instrument was calculated for this report that considers both the distance and the topography between a household and the "closest" (that is, cheapest-to-reach) market. Specifically, this variable, hereafter referred to as the "natural path," is the route that would be traveled if walking from a given location to the cheapest-to-reach market, in the absence of roads.[6] The natural path is thus the path that minimizes travel time, given the geography of the land. It therefore also represents the most cost-effective place to construct a road network, if economic benefits were ignored. Moreover, in the context of Africa it captures many of the historic trade and caravan routes where head-loading (walking) was the dominant precolonial mode of transportation. This instrument is an improvement over straight-line instruments given that the natural path more accurately represents what straight-line instruments are attempting to estimate, that is, the most logical (lowest-cost) route to connect two points, excluding considerations of benefits. Box 2.1 briefly discusses the value of the natural path as compared with the Euclidean distance IV. Annexes 2A and 2B to this chapter report results using both sets of IVs. The experiments indicate that the natural path and Euclidean distance are very similar in

Euclidean Distance versus Natural Path Instrumental Variable

In general the Euclidean distance and the natural path instrumental variables (IVs) produce very similar results. However, the difference between the two becomes important in areas where the topography of the land is very irregular, as would be the case in mountainous or swampy areas. In such cases, the natural path IV is expected to be more efficient because it takes such topography into account. (The topography of the land in between the production area and the market is what is being taken into account, not the topography of the land at the production area, which would significantly affect production and invalidate the natural path as a useful instrument.) To illustrate, consider the estimated impact of transport cost to market on local GDP in such regions of Nigeria. These regions were identified by (1) regressing Euclidean distance on the natural path and (2) identifying those observations with residuals in the top and bottom 2.5 percentiles. As shown in map B2.1.1, these areas correspond to the mountainous regions of Nigeria. Table B2.1.1 reports the local GDP estimates for these areas. Note that the ordinary least squares estimate has an upward (more negative) bias, which the IVs correct for. However, the Euclidean distance IV—which does not account for topography—does not go far enough in correcting this bias. These results are consistent with the natural path being more efficient than the Euclidean distance as an IV.

Map B2.1.1 Comparing Instrumental Variables over Varying Topography

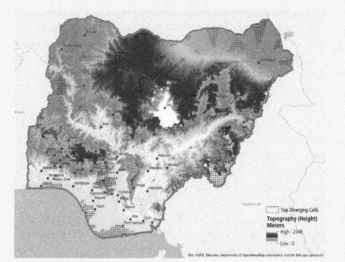

Source: Calculated from NASA's Shuttle Radar Topography Mission.

(continued next page)

Box 2.1 (continued)

Table B2.1.1 Euclidean Distance vs. Natural Path Instrumental Variable

	OLS	Euclidean distance IV	Natural path IV
ln(transport cost to market)	−0.539***	−0.386***	−0.260**
	(−32.86)	(−3.26)	(−2.04)
Observations	10,728	513	513
Other controls	Distance to mine, population and its square, low agricultural potential and its square (for cassava, yam, maize, and rice)		

Sources: Ghosh et al. 2010; and calculations.
Note: IV = instrumental variable; OLS = ordinary least squares. t-statistics in parentheses.
***$p < 0.01$, **$p < 0.05$, *$p < 0.10$.

topographies that are flat and uninterrupted, but differences emerge when there are impediments to traveling along a straight line. This finding suggests that the natural path is an improvement upon Euclidean distance in geographies with rugged terrain. In the interest of brevity, only the natural path IV estimates are discussed here.

In addition to handling these two classic issues using novel methodologies, this chapter attempts to improve upon previous efforts in two ways:

First, by using a number of different measures of well-being, the approach used here provides a more complete and robust picture of the extent to which wealth and incomes respond to a reduction in transport costs and whether there are significant differences. No single measure of welfare is perfect, but using a variety of metrics gives an indication of the robustness of conclusions. Welfare variables from two household surveys[7] are used along with a raster data set on local GDP derived from satellite images of nighttime lights (luminosity).[8] From the household surveys, several welfare indicators are calculated, including revenue from crop production, revenue from livestock sales, nonagricultural income, employment, asset wealth, and a multidimensional poverty indicator. Household survey data are useful because they allow specific heterogeneities of the households to be controlled for. The raster data set from Ghosh et al. (2010) provides an estimate of local GDP (proxied by luminosity) at a fine spatial level for the entire land area of Nigeria.

Using several measures also permits the study of the effects of transport costs on both "flow" and "stock" measures of welfare. Variables representing income flows (including crop revenue, livestock sales, and nonagricultural income) are used to give a snapshot of a family's well-being at a given time.

However, these flow measures of income may be highly affected by idiosyncratic, temporary, or localized shocks, for instance, a bad harvest due to less than average rainfall, or a sudden illness for the household head. In addition, although improving transportation infrastructure can lead to benefits in the short term, many of the benefits will not become apparent for many years after the improvement, after households and businesses have had time to adjust to a new equilibrium.[9] "Stock" measures of welfare can better capture the longer-term cumulative benefits than can the flow variables. In the long term, lower transport costs may produce multiplier effects that are unaccounted for in flow data—lower education costs, greater social cohesion, faster dispersion of production technologies, and so on—but that can increase productivity. Benefits from these road projects will not all occur immediately. They will likely cascade over time, as people begin learning of the new, lower transport costs and adjust their behavior accordingly. The more long-term indicators of welfare are analogous to the economic concept of "permanent income." For this reason, this chapter also studies the effects of road construction on two such variables: a wealth index available in the Nigeria Demographic and Health Survey (NDHS), and a multidimensional poverty index (MPI) that was constructed for this report, following Alkire and Santos (2010).[10] By looking at these different indicators of welfare, one can disaggregate the benefits of a transport cost reduction on total income into benefits on different components of income. This approach provides insights into the different aspects of economic activity, that is, agriculture, livestock, or nonagriculture, that are affected by transport cost reductions and thus improves the understanding of pathways through which transport cost reduction can lead to poverty reduction.

Furthermore, to explore how transport costs influence employment, two outcomes are considered: the probability of being employed year round and, conditional on being employed, the probability of being employed in agriculture. By considering these outcomes together, insight can be gained into whether roads increase economic activity thereby leading to greater employment opportunities and more diverse employment opportunities that allow workers to shift from the agricultural to the nonagricultural sector.

The final welfare measure used is local GDP, as constructed using nighttime lights imagery by Ghosh et al. (2010). Although the number of lights in an area is undoubtedly correlated with the GDP generated there, it is likely to be an even better measure of the stock of assets, since most of the variance in lighting among geographical areas is due to differences in degrees of electrification and in numbers of businesses and residential buildings, all of which require investments over time. In that sense, the lights variable is more akin to measures of wealth or permanent income than to measures of transitory income.

The second improvement is that the data on local GDP are exploited to develop a simple methodology for carrying out an evaluation of specific proposed road investments. Arguably, few if any other tools are available that would allow an intuitive, comprehensive, and rapid (for example, without detailed information on actual or potential traffic) comparative assessment to be made of options for road investments. In contrast to survey or census data, local GDP data are readily available for all areas, and thus allow for cross-regional analysis. Local GDP data provide baseline data on economic activity and allow for simulations to forecast benefits of proposed road projects at the regional level, as well as provide an additional result that is used to check the robustness of the household survey elasticities.

Estimating Impacts of Roads on Welfare

The main identification strategy for estimating benefits of existing roads is to instrument for transport cost to market with the natural path variable (that is, the time it takes to walk to market along the natural terrain). The details of the methodology are in box 2.2, and the estimated elasticities are summarized in Table 2.1. For comparison, the Euclidian distance IV results are reported in the tables in annex 2B. The estimated results using the two IVs are similar, and in the interest of brevity only the natural path IV results are discussed in this chapter. (See box 2.1 for a comparison of the natural path and the Euclidean distance, and an explanation of why the difference is important in certain areas.)

BOX 2.2

Technical Estimation

The regression analysis uses the following two-stage model:

$$\ln(Y_i^k) = \beta_0 + \beta_1 \ln(T_i) + X_i'\gamma + \varepsilon_i \tag{1}$$

$$\ln(T_i) = \alpha_0 + \alpha_1 \ln(N_i) + X_i'\theta + \mu_i \tag{2}$$

in which Y_i^k denotes the level of outcome k (agricultural revenue, livestock sales, nonagricultural income, employment, wealth index, and local GDP) indicating welfare of household i when using Living Standards Measurement Study–Integrated Surveys on Agriculture and Nigeria Demographic and Health Survey analysis or of location i when using nighttime lights data analysis. T_i is the transport cost to market, X_i is a vector of household and regional controls, and N_i is the natural path variable. The local GDP analysis includes geographic-level control variables, denoted by X_i, which are

(continued next page)

Box 2.2 (continued)

measured at the center of each raster cell. When analyzing the impact of transport costs on the probability of being multidimensionally poor, $dMPI_i$ would be the dependent variable in place of lnY_i, and $dMPI_i$ is a dummy taking the value of 1 if the household is multidimensionally poor and zero otherwise.

Equation (1) illustrates the impact of transport costs on the outcome of interest. Equation (2) shows the first-stage equation. The parameter of interest is β_1, which indicates the causal impact on household welfare of the cost of traveling to the closest market.

First, the regressions are run using an ordinary least squares (OLS) estimator with the transport cost included directly as an explanatory variable (that is, no instrumental variable [IV]). These results are then compared with the two-stage least squares procedure using the IV methodology described in the main text. The results yield the same sign, and most IV estimates do not differ greatly from the OLS, but given the expectation that these are biased as argued in the main text, they are not reported nor used in the following analysis.

As a robustness check on the IV estimates, a set of Conley bounds are calculated for the coefficient of interest; transport cost to market is calculated following Conley, Hansen, and Rossi (2012).

To illustrate the Conley bounds, consider the following model:

$$\ln(Y_i^k) = \beta_0 + \beta_1 \ln(T_i) + X_i'\gamma + \delta \ln(N_i) + \varepsilon_i. \qquad (3)$$

The traditional IV strategy assumes that $\delta = 0$. In other words, the variable N_i satisfies the exclusion restriction and can thus be excluded from the main equation. Conley, Hansen, and Rossi (2012) allow the parameter δ to be close to but not actually equal to zero; in other words, they allow the IV to be only "plausibly exogenous." The range of possible values for δ follows previous application of the Conley bound method in the literature (for example, Emran, Maret-Rakotondrazaka, and Smith 2014). By allowing the value of δ to vary, one can test the effect on the estimated parameter if the underlying assumption—that the IV is totally exogenous with respect to the dependent variable—is not entirely correct, that is, if there is some degree of endogeneity. This in effect tests the validity of instruments as well as estimates, and allows a plausible range for the estimated parameter to be calculated.

The Conley bound estimates indicate that even if the natural path instrument were not entirely exogenous, the estimates it produces would not be significantly biased, and the magnitude and direction of the estimated effects would not change significantly. (See annex 2B, table 2B.10 for Conley bound estimates.)

Table 2.1 Elasticities of Welfare Measure with Respect to Transport Cost

Welfare Indicator	Benefit (percent change resulting from a 10 percent reduction in transport costs)
Crop revenue	3.8
Livestock revenue	Negligible
Nonagricultural income	3.9
Year-round employment	0.4 (male), 0.3 (female)
Agricultural employment	−4.0 (male), −5.3 (female)
MPI reduction	2.6
Wealth index	2.0
Local GDP (from luminosity)	5.0

Note: MPI = multidimensional poverty index. These estimates are calculated using the natural path instrumental variable. Revenue figures are gross.

The estimates in table 2.1 are robust to a number of alternative specifications. For example, including geopolitical or agro-ecological zone dummies in place of marketshed fixed effects yields broadly consistent results. The LSMS results (crop revenue, livestock revenue, and nonagricultural income) are robust to the inclusion or exclusion of labor, land, fertilizer, and irrigation. Furthermore, alternative controls, such as availability of credit, as well as an indicator for the presence or absence of a hospital or school within the village, were tested.[11] Local GDP results are robust to the inclusion or exclusion of urban areas. Additionally, the gridded nature of the local GDP data exposes the analysis to the possibility for biased results caused by spatial autocorrelation. A spatial bootstrapping method is employed to test for this, and no bias is found.[12]

Although all of these variables are indicators of well-being, each one measures a somewhat different aspect of the impact of reducing transport cost. The first three elasticities show that the effect on crop production, on average, is higher than on livestock production, and that nonagricultural income is an important channel of transmission of benefits. Taken together, the increase in year-round employment and the decrease in agricultural employment—resulting from a decrease in transport costs—suggest that workers are switching from the agricultural to the nonagricultural sector. The measures of multidimensional poverty and wealth (assets) are of interest because they are more long term, that is, less prone to large changes from year to year. Furthermore, the local GDP data (a flow variable) are also a good proxy for local wealth, as argued previously in this chapter. In sum, the estimates obtained indicate that the welfare of the rural poor in Nigeria is clearly improved by the lowering of transport costs.

Using the Results to Evaluate the Economic Impact of Prospective Road Investments

In the final stage the potential welfare benefits from the completion of specific proposed road investments in Nigeria are evaluated using the estimated elasticity of local GDP with respect to transport cost. This calculation is made in a way that allows for heterogeneous benefits depending on current levels of welfare, current transport costs, and spatially varying transport cost reductions from road improvements. Furthermore, to account for spatial heterogeneity in the benefits from reduction in transport costs, new elasticities are calculated for each of Nigeria's six geopolitical zones. Although this methodology must be used with due awareness of its shortcomings (described below), it does provide decision makers with the capability to quantify the potential induced benefits of numerous options and to prioritize those that should be evaluated in more depth. This methodology could be used to study any road improvement or new road construction project in Nigeria and to prioritize investments. Further spatial disaggregation beyond regions is also feasible if needed for policy making. For instance, a separate elasticity could be calculated for each of Nigeria's 36 states and 1 territory. However, given the relatively smaller size of the states, this approach would result in a significant loss of statistical power.

To estimate the impact of improving the quality of a road segment, the portion of the New Economic Partnership for Africa's Development's (NEPAD's) and the African Development Bank's proposed Trans African Highways Project segments that run through Nigeria were selected as case studies: (1) a north-south corridor that connects Niger (going through Kano) and Lagos (1,120 kilometers); (2) a northeastern corridor that connects Niger and Cameroon (940 kilometers); and (3) a southeastern corridor that connects Lagos with Cameroon (730 kilometers).[13]

Methodology
To calculate transport costs for any hypothetical road project, first the project is digitized, and it is assumed that each corridor would be improved from its current quality to paved and good condition status. The baseline scenario (current quality) is obtained from FERMA, and although this information shows that only 20 kilometers need to be paved, approximately 1,275 kilometers need to be improved from poor to good and 815 kilometers from fair to good condition.

To calculate the change in transport cost due to the improvement of each corridor, the same procedure that was used to estimate the travel cost to the closest market is again followed. That is, for each of the three NEPAD projects the optimal route to the "nearest" market is reestimated following the same procedure, thereby obtaining the cheapest travel cost for every cell, assuming that each of these roads is entirely in good, paved condition. Then the results are compared with the baseline scenario to obtain the change in transport cost due to each road-improvement project, for each grid cell. The economic agents (households) analyzed in this chapter spatially interact with markets only at small distances. Hence, the local market was selected as the main destination. It is worth noting that the network analysis to find the best route for each agent uses tertiary, secondary, and primary roads. Thus, the analysis is capturing rural connectivity and market issues at the same time. The percentage change in transport costs for each cell, if all three of the corridor improvement projects were to be completed, is shown in map 2.2.

The increase in local GDP is estimated in each grid cell separately, and then aggregated to arrive at the total benefit. A different transport cost elasticity is calculated for each of Nigeria's six geopolitical zones.[14] These estimates are reported in table 2.2. In each grid cell, the relevant elasticity is multiplied by the percentage change in transport costs due to each road construction project to arrive at the percentage increase in local GDP for that grid cell. That percentage increase in local GDP is then multiplied by the baseline level of GDP (from the local GDP data) to generate the actual increase in local GDP for that grid cell. This increase in local GDP is then summed across all grid cells to arrive at an aggregate value.[15]

Map 2.2 Percentage Change in Transport Costs

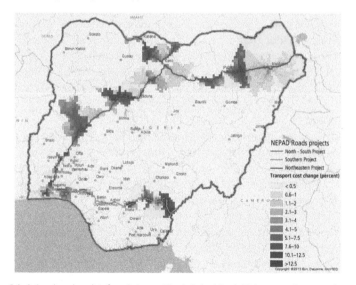

Sources: Calculations based on data from DeLorme; Nigeria Federal Roads Maintenance Agency; and World Bank 2013.
Note: NEPAD = New Economic Partnership for Africa's Development.

Table 2.2 Transport Cost Elasticities by Region

Region	Transport cost elasticities	Upper bound of 95 percent confidence interval	Lower bound of 95 percent confidence interval
South East	−1.095***	−1.275	−0.914
	(−11.89)		
South West	−0.534***	−0.738	−0.330
	(−5.14)		
South South	−0.710***	−0.910	−0.510
	(−6.97)		
North East	−0.273**	−0.462	−0.084
	(−2.82)		
North West	−0.562***	−0.754	−0.370
	(−5.73)		
North Central	−0.549***	−0.736	−0.362
	(−5.74)		

Sources: Ghosh et al. 2010; and calculations.
Note: t-statistics in parentheses.
***$p < 0.01$, **$p < 0.05$, *$p < 0.10$.

This approach also allows the total number of Nigerians who would benefit from each road construction project to be estimated, and an efficiency analysis to be conducted, by estimating the benefit per road-kilometer improved. Given the inherent uncertainty involved in statistical analysis, the total benefits are calculated using the point estimates of elasticity as well as a range of benefits representing a 95 percent confidence interval (in table 2.2).

Point estimates of benefits
The transport cost reduction obtained from each road construction project is estimated separately, and then the total benefits if all of the projects were completed is estimated. Note that the total benefits of all of the construction projects is not equal to the sum of the benefits of each of the projects individually because there is some overlap among the projects' locations. New transport cost elasticities are then calculated, one for each of Nigeria's six geopolitical zones, by interacting geopolitical zone fixed effects with market transport costs.

Table 2.3 presents the point estimates for the benefits of the three NEPAD projects analyzed. The north-south corridor, which is the longest road of the three projects, would result in estimated annual benefits of more than $1 billion. Annual benefits from the northeastern and southern corridors are significantly lower, at $233 million and $529 million, respectively. Nevertheless, these roads are also shorter, potentially implying lower costs of improvement. If all projects were completed, total estimated annual benefits would be $1.5 billion. Map 2.3 shows the distribution of these benefits.

Table 2.3 NEPAD Project Benefits

	GDP increase (million US$, 2006 PPP)	Road length (km)	GDP increase per km (million US$)	Population affected (millions)	Per capita benefit (US$ per person affected)
All projects	1,794	2,774	0.65	39.32	45.63
North-south corridor	1,082	1,121	0.97	23.12	46.81
Northeastern corridor	233	939	0.25	14.88	15.66
Southern corridor	529	729	0.73	9.16	57.73

Source: World Bank calculations.
Note: PPP = purchasing power parity.

Map 2.3 Increase in Local GDP
(US$ million per 100 square kilometer cell)

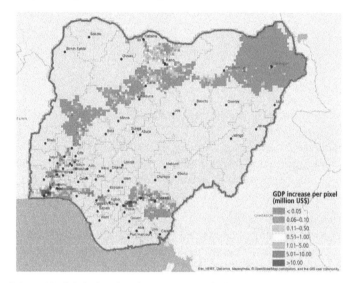

Sources: DeLorme; Nigeria Federal Roads Maintenance Agency; Ghosh et al. 2010; World Bank 2013; and calculations.

Notice that the third column of table 2.3, which shows the total benefit per kilometer of each corridor, suggests that projects can be ranked according to their benefit-efficiency. That is, assuming road improvement costs per kilometer are uniform and equal across projects, the project that provides the most benefits per unit cost can be discerned.[16] The north-south corridor project would return annual benefits of approximately $970,000 per kilometer improved, significantly higher than that of the northeastern and southern corridors, which have benefits of $250,000 and $730,000 per kilometer improved, respectively. Map 2.4 visually displays the benefits per kilometer of each road segment.

Map 2.4 Benefits per Kilometer of Road Improved (US$ million)

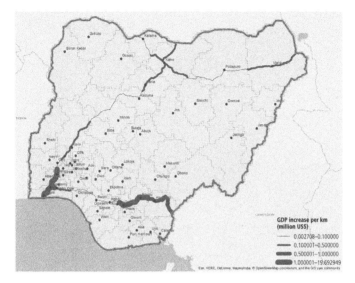

Sources: DeLorme; Nigeria Federal Roads Maintenance Agency; World Bank 2013; and calculations.

Using LandScan (2006) population data, it is possible to obtain an estimate of the number of people who would be affected by each project. Again, the north-south corridor has the biggest impact, benefiting 23.1 million people, as a result of its length. The northeastern and southern corridor projects would benefit 14.9 million and 9.2 million people, respectively. Map 2.5 shows the total population affected if all three corridor improvement projects were to be completed. Dividing total benefits by the number of people affected yields estimated benefits per person affected. The north-south corridor project, in addition to benefiting the largest number of people, also has the biggest benefit per capita, at $45.6 per person benefited. Map 2.6 visually displays the population benefited per kilometer of each road segment.

Road segment prioritization
Decision makers would find it useful to know what segments have the highest benefits, so they can set priorities. Therefore, as a final exercise, the segments of each road are analyzed separately. Of course, each of these road projects is not all or nothing; portions could be improved sequentially over time, and some could not be improved at all.

The analysis that follows first splits up the roads into segments, each of which services one "marketshed," defined as the area around each city (with a population of at least 100,000 residents) from which the residents can travel to that city

Map 2.5 Population Affected by Corridor Improvement Projects

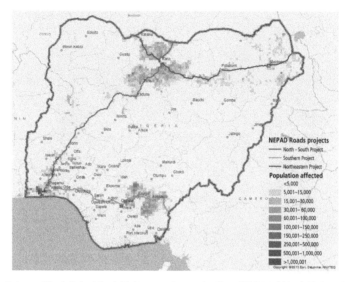

Sources: DeLorme; Nigeria Federal Roads Maintenance Agency; LandScan 2006; World Bank 2013; and calculations.

at lower transport costs than to any other city. The size and shape of a city's marketshed depends on both the road network around that city and its proximity to other cities. Each road is partitioned into segments according to the marketshed that uses each segment. This is, by construction, a model of local travel; that is, there is no travel between marketsheds using the road. The estimates based on this methodology are therefore conservative, given that people will likely also derive some benefits from improvements to roads that mainly benefit marketsheds outside those in which they live.

Table 2.4 shows a list of the marketsheds described above that use a portion of one of the three NEPAD roads and would therefore benefit from its improvement. Columns 3–8 quantify benefits according to several criteria a policy maker might use to prioritize each segment. By most of these criteria, the road segment benefiting the Lagos marketshed would produce the greatest benefits.

Robustness Checks

Given that statistical estimates are not precise, a plausible range of benefits is presented based on the 95 percent confidence interval surrounding the estimated local GDP elasticity of transport costs by region. These intervals are given in table 2.2. Using these elasticities to recalculate total annual estimated benefits for each of the NEPAD transport improvement projects yields the values in table 2.5. If all projects were completed, benefits would range between $1.206 billion and $2.383 billion.

Table 2.4 Road Improvement Prioritization

Marketshed	(1) Length of road improved (km)	(2) Total GDP (million US$ PPP)	(3) GDP increase (million US$, PPP)	(4) Percentage increase in GDP (3)/(2)	(5) Total population	(6) Population affected	(7) GDP increase per km improved (million US$/km) (3)/(1)	(8) Population affected per km of road (6)/(1)
Abakaliki	203.1	3,087	114.52	3.71	3,592,510	3,587,821	0.564	17,662
Bauchi	56.8	2,808	0.21	0.01	2,503,664	135,459	0.004	2,387
Bida	30.7	1,607	0.91	0.06	1,180,780	84,630	0.030	2,760
Enugu	90.4	7,556	133.43	1.77	3,682,892	2,705,968	1.476	29,929
Ibadan	87.1	7,832	72.04	0.92	2,587,179	1,219,364	0.827	13,999
Ijebu Ode	87.6	2,844	32.18	1.13	788,070	576,516	0.367	6,581
Ilorin	199.8	4,682	59.80	1.28	2,159,909	659,174	0.299	3,299
Kaduna	180.8	9,512	55.59	0.58	2,056,143	632,083	0.307	3,496
Kano	212.0	17,652	98.66	0.56	13,158,975	7,488,524	0.465	35,324
Katsina	118.7	5,410	80.88	1.49	4,351,990	1,485,371	0.681	12,509
Lagos	34.2	41,860	681.38	1.63	10,413,453	8,618,237	19.900	251,703
Maiduguri	30.6	4,787	37.12	0.78	3,836,727	3,075,144	1.214	100,536
Minna	132.9	4,617	5.97	0.13	1,780,793	187,871	0.045	1,413
Ogbomosho	47.0	887	21.43	2.42	663,362	326,754	0.456	6,955
Okitipupa	69.3	1,201	1.37	0.11	955,233	127,347	0.020	1,838
Onitsha	40.1	7,446	238.83	3.21	2,778,550	1,935,824	5.952	48,247
Oyo	28.0	844	5.07	0.60	528,668	203,382	0.181	7,264
Potiskum	265.6	3,807	24.49	0.64	4,950,803	1,544,431	0.092	5,815
Shagamu	76.6	4,042	100.51	2.49	653,758	653,758	1.312	8,533
Zaria	97.6	5,089	24.49	0.48	4,496,220	3,056,001	0.251	31,308

Source: World Bank calculations.
Note: PPP = purchasing power parity.

Table 2.5 NEPAD Project Benefits, 95 Percent Confidence Interval
(million US$ per kilometer of road improved)

	GDP increase, lower bound	GDP increase, point estimate	GDP increase, upper bound
All projects	1,206	1,794	2,383
North-south corridor	678	1,082	1,487
Northeastern corridor	132	233	334
Southern corridor	429	529	629

Source: World Bank calculations.

Map 2.6 Population Benefited per Kilometer Improved

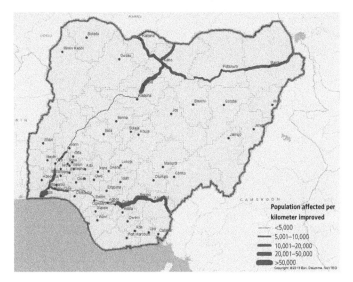

Sources: DeLorme; Nigeria Federal Roads Maintenance Agency; LandScan 2006; World Bank 2013; and calculations.

Final Words and Caveats

The analysis presented in this chapter demonstrates a robust method for estimating the economic impact from several road improvement projects. Although
the best possible methods and the best possible data are used, several shortcomings must be acknowledged.

There are uncertainties. In estimating the elasticity of local GDP with respect
to transport cost, the data used for the dependent variable are not observable
directly, but rather are estimated using nighttime lights. This approach adds an
additional level of uncertainty to the estimates, but it is an uncertainty that is
unavoidable given that spatially disaggregated data on actual (nonestimated) GDP

are not available. However, even though this elasticity is based on estimated data, it is similar in magnitude to elasticities estimated using other income variables directly observed from survey data, providing some confidence in its robustness.

The estimated effects on local GDP are best interpreted as partial equilibrium results. However, under the assumption that the reduced-form estimate of local GDP provides an accurate proxy for the true generating process, it could be argued that the simulated impacts using local GDP might serve as a proxy for some general equilibrium effects. In other words, the methodology used might capture general equilibrium effects but only in the area close to the road, not economywide. Effectively capturing the general equilibrium effects would require estimating a structural model of the entire (or relevant parts of the) economy. Because of data limitations, such an exercise is beyond the scope of this report.

When interpreting the estimated elasticities, a few caveats must be kept in mind. First, the revenue outcome variables analyzed (from crops, livestock, and nonagricultural sources) are gross. Inputs, such as labor and land, are controlled for, but because of market imperfections, determining the associated costs is often challenging. This difficulty may, in part, explain the relatively high estimates as compared with other examples in the literature, most of which report net impacts. Furthermore, the estimated elasticities reported in the literature are subject to significant endogeneity. See, for example, Dorosh et al. (2012), which finds even larger elasticities, or Ulimwengu et al. (2009) for the Democratic Republic of Congo.

The estimated elasticities are robust to a number of alternative specifications, including controlling for regional (agro-ecological zone or geopolitical) dummies in place of marketshed fixed effects. The revenue outcomes are robust to the inclusion or exclusion of land, labor, fertilizer, or irrigation controls. Several additional control variables were tried on the right-hand side, including access to credit and presence of hospitals or schools.

This report focuses on the impact of road infrastructure. A comparison of the relative benefits of alternative investments (for example, in education, railways, communication technology), though interesting, is left for future research.

Another potential shortcoming is that cross-sectional data are used, which can often make discerning the direction of causality very difficult. The IV technique used here is one very commonly used in the literature, and it is argued that the natural path IV is a significant improvement over those used in other widely cited research. Nevertheless, there is no such thing as a perfect instrument. For this reason, the point estimates are presented along with the respective Conley bounds, which give a range of estimates under the assumption that the instrument is not perfectly exogenous (shown in annex 2B, table 2B.10). The ranges given by the Conley bound estimates are relatively small, showing that even if the natural path instrument were to violate the necessary exclusion restriction, the point estimates would not be dramatically wrong.

Finally, any evaluation of infrastructure projects must, in addition to the economic benefits, as estimated here, account for potential social and environmental costs, such as displacement effects and externalities. Some of these additional benefits and costs are explored in subsequent chapters. The next chapter investigates one potential mechanism that could explain the positive economic impact, specifically, how a reduction in transport costs induces greater adoption of modern technology in agriculture and how this in turn leads to improved well-being.

Annex 2A Summary Statistics

Table 2A.1 Summary Statistics from the Nigeria Living Standards Measurement Study

Variable	Observations	Mean	Standard deviation	Minimum	Maximum
Outcome					
Crop revenue (US$)	2,600	157.84	2,126.75	0.00	105,600.00
Non-agriculture income (US$)	2,600	1,281.07	35,930.45	0.00	1,320,013
Non-agriculture income (US$)	2,600	0.37	0.15	0.00	0.83
Total sales of livestock (US$)	3,297	48.77	204.33	0.00	3,960.00
Treatment					
Cost of transporting one ton of goods to market (US$)	2,600	5.10	3.30	0.14	17.36
Instrumental variable					
Natural path to market (hours)	2,600	14.06	10.43	0.00	59.42
Controls					
Age of household head	2,600	51.40	15.10	15.00	110.00
Dummy: Household head is literate	2,600	0.55	0.50	0.00	1.00
Land (km²)	2,600	9.23	16.92	0.00	265.03
Number of workers in the house	2,600	2.92	2.14	0.00	18.00
Household members working on own plot	2,600	2.02	2.03	0.00	17.00
Total fertilizer used (kg)	2,600	11.14	79.91	0.00	2,299.00
Dummy: Household irrigates its plot	2,600	0.95	0.21	0.00	1.00
Dummy: Tropical warm semi-arid zone	2,600	0.31	0.46	0.00	1.00
Dummy: Tropical warm sub-humid zone	2,600	0.59	0.49	0.00	1.00
Dummy: Tropical warm humid zone	2,600	0.09	0.28	0.00	1.00
Dummy: Tropical cool sub-humid zone	2,600	0.01	0.11	0.00	1.00
Dummy: North East Region	2,600	0.17	0.38	0.00	1.00
Dummy: North West Region	2,600	0.19	0.40	0.00	1.00
Dummy: South East Region	2,600	0.23	0.42	0.00	1.00
Dummy: South South Region	2,600	0.14	0.35	0.00	1.00

(continued next page)

Table 2A.1 (continued)

Variable	Observations	Mean	Standard deviation	Minimum	Maximum
Dummy: South West Region	2,600	0.09	0.29	0.00	1.00
Total household business expenses (US$)	2,600	613.69	25,888.87	0.00	1,320,000
Costs of livestock (US$)	3,297	58.90	313.26	0.00	10,312.50

Sources: Nigeria Living Standards Measurement Study–Integrated Surveys on Agriculture (National Bureau of Statistics 2010); and calculations.

Table 2A.2 Summary Statistics from the Nigeria Demographic and Health Survey

	Mean	Standard deviation	Minimum	Maximum
Outcomes				
Wealth index	−12,407	98,657.2	−145,026	305,274
Multidimensionally poor (dummy)	0.703	0.457	0	1
Year-round employment, male (dummy)	0.534	0.499	0	1
Year-round employment, female (dummy)	0.426	0.495	0	1
Agricultural employment, male (dummy)	0.466	0.499	0	1
Agricultural employment, female (dummy)	0.283	0.451	0	1
Variable of interest				
Transport cost to market (US$)	5.652	4.097	0.290	30.301
Instruments				
Time taken to reach market using natural path (hours)	15.185	12.417	0	70.794
Controls				
Agricultural potential (factor of ln agricultural potential for cassava, maize, and rice)	−0.043	0.966	−1.554	1.211
Dummy: Household agriculturally involved	0.732	0.443	0	1
Age of household head	40.031	11.268	17	99
Sex of household head	1.034	0.182	1	2
Number of household members	6.517	3.079	3	43
Number of female members in households age 15–49 years	1.418	0.775	1	15
Number of male members in households age 15–59 years	1.279	0.678	1	12
Number of children age 0 to 5 years	1.706	0.860	1	9
Dummy: Residence is rural	0.713	0.452	0	1
Dummy: North East Region	0.216	0.412	0	1
Dummy: North West Region	0.271	0.445	0	1
Dummy: South East Region	0.080	0.271	0	1
Dummy: South South Region	0.114	0.318	0	1
Dummy: South West Region	0.130	0.336	0	1
Number of observations: 6,684				

Sources: Nigeria Demographic and Health Survey (National Population Commission 2009); and calculations.

Table 2A.3 Summary Statistics from Local GDP Data Sets

	Mean	Standard Deviation	Minimum	Maximum	Label
Local GDP	25.369	145.317	0	4,469.6	Total income per cell (million US$, 2006 PPP)
Population	13.64	44.450	0	1,639.2	Population per cell (thousands), LandScan 2006
Cassava potential yield	833.5371	681.500	0	2,775	Yield (kg/ha), FAO and IIASA 2000
Rice potential yield	494.7672	436.507	0	1,792	Yield (kg/ha), FAO and IIASA 2000
Yams potential yield	609.166	383.281	0	1,747	Yield (kg/ha), FAO and IIASA 2000
Maize potential yield	1,209.439	625.555	0	3,556	Yield (kg/ha), FAO and IIASA 2000
Number of observations: 10,015					

Sources: Ghosh et al. 2010; LandScan 2006; FAO and IIASA 2000; and calculations.
Note: kg/ha = kilograms per hectare; PPP = purchasing power parity.

Annex 2B Regression Results

Table 2B.1 Crop Revenue

Dependent variable: ln(crop revenue)	(1) Ordinary least squares	(2) Instrumental variable, Euclidean distance	(3) Instrumental variable, natural path
ln(transport cost to market)	−0.827***	−0.849***	−0.970***
	(−5.40)	(−5.16)	(−4.83)
Household agricultural labor	0.159**	0.161**	0.170**
	(2.48)	(2.51)	(2.56)
Household agricultural labor^2	−0.020***	−0.020***	−0.021***
	(−3.40)	(−3.42)	(−3.43)
Land	0.017***	0.017***	0.018***
	(4.09)	(4.12)	(4.19)
Fertilizer	0.001	0.001	0.001*
	(1.63)	(1.64)	(1.67)
Dummy = 1 if irrigates land	0.487	0.489	0.497
	(1.15)	(1.16)	(1.20)
Dummy = 1 if tropical warm subhumid	0.346*	0.341*	0.312*
	(1.89)	(1.86)	(1.72)
Dummy = 1 if tropical warm humid	−0.022	−0.029	−0.072
	(−0.08)	(−0.11)	(−0.26)
Dummy = 1 if tropical cool humid	1.719***	1.711***	1.667***
	(7.94)	(7.88)	(7.48)

(continued next page)

Table 2B.1 (continued)

Dependent variable: ln(crop revenue)	(1) Ordinary least squares	(2) Instrumental variable, Euclidean distance	(3) Instrumental variable, natural path
Age of household head	0.014	0.014	0.013
	(0.87)	(0.85)	(0.77)
Age^2	−0.000	−0.000	−0.000
	(−0.72)	(−0.71)	(−0.65)
Dummy = 1 if household head is literate	0.312***	0.309***	0.298***
	(2.84)	(2.82)	(2.65)
Constant	1.785***	1.830***	2.090***
	(3.16)	(3.09)	(3.31)
First-stage results			
Instrumental variable: ln(natural path)		0.603***	0.631***
		(38.78)	(19.28)
Angrist-Pischke test of weak identification		1,504.10	371.66
		P = 0.0000	P = 0.0000
Observations	2,600	2,600	2,600

Sources: Nigeria Living Standards Measurement Study–Integrated Surveys on Agriculture (National Bureau of Statistics 2010); and calculations.
Note: Robust t-statistics clustered at the enumeration area in parentheses.
***p < 0.01, **p < 0.05, *p < 0.10.

Table 2B.2 Livestock Sales

Dependent variable: ln(livestock sales)	(1) Ordinary least squares	(2) Instrumental variable, Euclidean distance	(3) Instrumental variable, natural path
ln(transport cost to market)	−0.026	−0.140	−0.161
	(−0.15)	(−0.76)	(−0.87)
ln(cost of animals)	0.186***	0.186***	0.186***
	(3.85)	(3.88)	(3.88)
Household agricultural labor	0.056	0.060	0.061
	(0.65)	(0.71)	(0.72)
Household agricultural labor^2	−0.001	−0.001	−0.001
	(−0.14)	(−0.16)	(−0.16)
Land	0.008**	0.009**	0.009**
	(2.19)	(2.36)	(2.39)
Fertilizer	−0.000	−0.000	−0.000
	(−0.98)	(−0.92)	(−0.91)

(continued next page)

Table 2B.2 (continued)

Dependent variable: ln(livestock sales)	(1) Ordinary least squares	(2) Instrumental variable, Euclidean distance	(3) Instrumental variable, natural path
Dummy = 1 if irrigates land	0.461	0.465	0.466
	(1.36)	(1.38)	(1.38)
Dummy = 1 if tropical warm subhumid	0.355*	0.336*	0.332*
	(1.90)	(1.80)	(1.77)
Dummy = 1 if tropical warm humid	−0.134	−0.185	−0.195
	(−0.47)	(−0.65)	(−0.68)
Dummy = 1 if tropical cool humid	−0.245	−0.274	−0.280
	(−1.23)	(−1.39)	(−1.42)
Age of household head	0.027	0.027	0.027
	(1.43)	(1.40)	(1.40)
Age^2	−0.000*	−0.000*	−0.000*
	(−1.72)	(−1.72)	(−1.72)
Dummy = 1 if household head is literate	−0.317**	−0.326***	−0.328***
	(−2.49)	(−2.61)	(−2.63)
Constant	0.116	0.342	0.385
	(0.17)	(0.50)	(0.56)
First-stage results			
Instrumental variable: ln(natural path)		0.606***	0.649***
		(28.28)	(25.46)
Angrist-Pischke test of weak identification		799.67	648.20
		P = 0.0000	P = 0.0000
Observations	3,297	3,297	3,297

Sources: Nigeria Living Standards Measurement Study–Integrated Surveys on Agriculture (National Bureau of Statistics 2010); and calculations.
Note: Robust t-statistics clustered at the enumeration area in parentheses.
***p < 0.01, **p < 0.05, *p < 0.10.

Table 2B.3 Nonagricultural Income

Dependent variable: Nonagricultural Income	(1) Ordinary least squares	(2) Instrumental variable, Euclidean distance	(3) Instrumental variable, natural path
ln(transport cost to market)	−0.421***	−0.461***	−0.464***
	(−3.35)	(−3.39)	(−3.38)
Age of household head	−0.052***	−0.052***	−0.052***
	(−2.62)	(−2.63)	(−2.63)

(continued next page)

Table 2B.3 (continued)

Dependent variable: Nonagricultural Income	(1) Ordinary least squares	(2) Instrumental variable, Euclidean distance	(3) Instrumental variable, natural path
Age^2	0.000**	0.000**	0.000**
	(2.47)	(2.47)	(2.47)
Dummy = 1 if household head is literate	0.376***	0.373***	0.373***
	(3.48)	(3.46)	(3.46)
Land	0.003	0.003	0.003
	(0.76)	(0.84)	(0.85)
Household labor	0.287***	0.288***	0.288***
	(3.04)	(3.07)	(3.07)
Household labor^2	−0.015	−0.015*	−0.015*
	(−1.64)	(−1.65)	(−1.65)
Dummy = 1 if North East	−0.503**	−0.504**	−0.504**
	(−2.27)	(−2.29)	(−2.29)
Dummy = 1 if North West	−0.266	−0.262	−0.262
	(−1.23)	(−1.22)	(−1.22)
Dummy = 1 if South East	−0.236	−0.250	−0.251
	(−0.99)	(−1.06)	(−1.06)
Dummy = 1 if South South	0.824***	0.812***	0.811***
	(3.69)	(3.67)	(3.66)
Dummy = 1 if South West	−0.039	−0.062	−0.064
	(−0.15)	(−0.24)	(−0.24)
Total business expenses	0.000***	0.000***	0.000***
	(28.29)	(28.47)	(28.47)
Constant	5.191***	5.261***	5.266***
	(9.49)	(9.39)	(9.35)
First-stage results			
Instrumental variable: ln(natural path)		0.595***	0.638***
		(32.46)	(19.89)
Angrist-Pischke test of weak identification		1,053.94	395.48
		P = 0.0000	P = 0.0000
Observations	1,355	1,355	1,355

Sources: Nigeria Living Standards Measurement Study–Integrated Surveys on Agriculture (National Bureau of Statistics 2010); and calculations.
Note: Robust *t*-statistics clustered at the enumeration area in parentheses.
***$p < 0.01$, **$p < 0.05$, *$p < 0.10$.

Table 2B.4 Year-Round Employment

Dependent variable: Dummy = 1 if employed year round and 0 if not employed or seasonally employed	Male			Female		
	(1)	(2)	(3)	(4)	(5)	(6)
	Ordinary least squares	Instrumental variable, Euclidean distance	Instrumental variable, natural path	Ordinary least squares	Instrumental variable, Euclidean distance	Instrumental variable, natural path
ln(transport cost to market)	−0.0430***	−0.0374**	−0.0407**	−0.0442***	−0.0324***	−0.0300***
	(−2.971)	(−2.344)	(−2.281)	(−5.427)	(−3.517)	(−2.979)
Agricultural potential	−0.0209*	−0.0209*	−0.0209*	−0.0101	−0.00858	−0.00861
	(−1.851)	(−1.861)	(−1.858)	(−1.634)	(−1.372)	(−1.376)
ln(age)	0.490***	0.490***	0.490***	0.491***	0.491***	0.491***
	(17.78)	(17.82)	(17.81)	(32.14)	(32.24)	(32.22)
Education level: Primary	0.156***	0.157***	0.157***	0.0718***	0.0714***	0.0719***
	(4.699)	(4.736)	(4.720)	(5.461)	(5.589)	(5.636)
Education level: Secondary	0.0798**	0.0815**	0.0805**	0.00451	0.00572	0.00659
	(2.339)	(2.385)	(2.353)	(0.291)	(0.386)	(0.444)
Education level: Higher than secondary	0.0373	0.0394	0.0381	−0.00499	−0.00245	−0.00128
	(0.921)	(0.974)	(0.939)	(−0.237)	(−0.117)	(−0.0616)
Dummy = 1 if rural	−0.0484*	−0.0521**	−0.0500*	−0.0151	−0.0149	−0.0165
	(−1.898)	(−2.018)	(−1.889)	(−1.078)	(−1.051)	(−1.132)
First-stage results						
Instrumental variable: ln(Euclidean distance)		0.791***			0.790***	
		(70.310)			(70.360)	
Instrumental variable: ln(natural path)			0.786***			0.793***
			(19.030)			(19.620)
Angrist-Pischke test of weak identification		4,943.25	362.16		4,951.01	384.89
		P = 0.0000	P = 0.000		P = 0.000	P = 0.000
Observations	33,918	33,918	33,918	36,703	36,703	36,703

Sources: Nigeria Demographic and Health Survey (National Population Commission 2009); and calculations.
Note: Robust *t*-statistics clustered at the enumeration area in parentheses. Regressions also control for religion, ethnicity, and marketshed fixed effects.
***p < 0.01, **p < 0.05, *p < 0.10.

Table 2B.5 Agricultural Employment

Dependent variable: Agricultural employment among employed individuals	Male			Female		
	(1)	(2)	(3)	(4)	(5)	(6)
	Probit	Instrumental variable Probit, Euclidean distance	Instrumental variable Probit, natural path	Probit	Instrumental variable Probit, Euclidean distance	Instrumental variable Probit, natural path
ln(transport cost to market)	0.095***	0.379***	0.403***	0.087***	0.425***	0.528***
	(7.140)	(5.920)	(5.910)	(7.710)	(7.040)	(7.610)
Agricultural potential	0.037***	0.163***	0.161***	0.012	0.065	0.062
	(3.010)	(3.080)	(3.030)	(1.550)	(1.580)	(1.510)
ln(age)	0.024	0.100	0.102	0.032***	0.167***	0.169***
	(1.000)	(0.980)	(1.000)	(2.220)	(2.240)	(2.270)
Education level: Primary	−0.149***	−0.571***	−0.567***	−0.090***	−0.402***	−0.390***
	(−5.19)	(−5.38)	(−5.33)	(−5.56)	(−5.60)	(−5.42)
Education level: Secondary	−0.262***	−1.001***	−0.996***	−0.192***	−0.915***	−0.894***
	(−8.73)	(−9.10)	(−9.06)	(−11.55)	(−12.24)	(−11.83)
Education level: Higher than secondary	−0.401***	−1.614***	−1.610***	−0.323***	−1.950***	−1.910***
	(−11.48)	(−10.27)	(−10.25)	(−18.57)	(−16.37)	(−15.79)
Dummy = 1 if rural	0.205***	0.852***	0.844***	0.135***	0.730***	0.679***
	(7.780)	(7.950)	(7.790)	(7.280)	(7.210)	(6.400)
First-stage results						
Instrumental variable: ln(Euclidean distance)		0.787***			0.786***	
		(71.300)			(63.030)	
Instrumental variable: ln(natural path)			0.828***			0.781***
			(29.910)			(17.080)
Observations	28,629	28,629	28,629	21,656	21,656	21,656

Sources: Nigeria Demographic and Health Survey (National Population Commission 2009); and calculations.
Note: Robust *t*-statistics clustered at the enumeration area in parentheses. Regressions also control for religion, ethnicity, and marketshed fixed effects.
***p < 0.01, **p < 0.05, *p < 0.10.

Table 2B.6 Wealth Index

Dependent variable: ln(wealth index)	(1) Ordinary least squares	(2) Instrumental variable, Euclidean distance	(3) Instrumental variable, natural path
ln(transport cost to market)	−0.235***	−0.210***	−0.204***
	(−10.280)	(−8.654)	(−7.758)
Agricultural potential	−0.0152	−0.0159	−0.0161
	(−0.878)	(−0.922)	(−0.930)
Dummy = 1 if household agriculturally involved	−0.235***	−0.244***	−0.246***
	(−8.451)	(−8.746)	(−8.828)
ln(age of household head)	−0.105***	−0.103***	−0.103***
	(−3.233)	(−3.187)	(−3.176)
Female household head dummy	−0.0189	−0.0207	−0.0211
	(−0.498)	(−0.544)	(−0.554)
ln(number of household members)	0.0574*	0.0563*	0.0560*
	(1.831)	(1.805)	(1.796)
ln(number of females age 15–49 years)	0.0903***	0.0910***	0.0912***
	(3.799)	(3.847)	(3.851)
ln(number of males age 15 to 59 years)	0.0521**	0.0543**	0.0548**
	(2.279)	(2.385)	(2.411)
ln(number of children age 0 to 5 years)	−0.0266	−0.0258	−0.0256
	(−1.449)	(−1.412)	(−1.400)
Dummy = 1 if rural	−0.441***	−0.457***	−0.461***
	(−10.46)	(−10.92)	(−10.88)
Constant	12.97***	12.97***	12.97***
	(109.8)	(109.9)	(109.8)
First-stage results			
Instrumental variable: ln(Euclidean distance)		0.791***	
		(65.870)	
Instrumental variable: ln(natural path)			0.852***
			(33.030)
Angrist-Pischke test of weak identification		4,338.48	1,091.24
		P = 0.0000	P = 0.0000
Observations	6,684	6,684	6,684

Sources: Nigeria Demographic and Health Survey (National Population Commission 2009); and calculations.
Note: Robust *t*-statistics clustered at the enumeration area in parentheses.
***$p < 0.01$, **$p < 0.05$, *$p < 0.10$.

Table 2B.7 Multidimensional Poverty, Nigeria Demographic and Health Survey

Dependent variable: Dummy = 1 if poor	(1) Ordinary least squares	(2) Instrumental variable, Euclidean distance	(3) Instrumental variable, natural path	(4) Probit	(5) Instrumental variable, probit Euclidean distance	(6) Instrumental variable, probit natural path
ln(transport cost to market)	0.0865***	0.0769***	0.0669***	0.078***	0.304***	0.262***
	(6.940)	(5.843)	(4.823)	(11.840)	(9.440)	(7.320)
Agricultural potential	0.00934	0.00962	0.00991	0.007	0.036	0.036
	(0.984)	(1.008)	(1.031)	(1.230)	(1.340)	(1.370)
Dummy = 1 if household agriculturally involved	0.124***	0.128***	0.132***	0.0932***	0.421***	0.437***
	(6.983)	(7.158)	(7.311)	(8.020)	(8.080)	(8.350)
ln(age of household head)	−0.0182	−0.0190	−0.0197	−0.017	−0.077	−0.079
	(−0.774)	(−0.804)	(−0.835)	(0.770)	(−0.78)	(−0.8)
Female household head dummy	0.0627*	0.0634*	0.0641*	0.036	0.16	0.167
	(1.843)	(1.854)	(1.867)	(1.500)	(1.460)	(1.510)
ln(number of household members)	0.0321	0.0325	0.0329	0.033	0.143	0.144
	(1.443)	(1.461)	(1.479)	(1.570)	(1.560)	(1.570)
ln(number of females age 15–49 years)	0.114***	0.113***	0.113***	0.112***	0.492***	0.489***
	(6.656)	(6.636)	(6.613)	(6.800)	(6.740)	(6.700)
ln(number of males age 15–59 years)	0.0229	0.0221	0.0212	0.014	0.057	0.056
	(1.253)	(1.204)	(1.154)	(0.820)	(0.770)	(0.760)
ln(number of children age 0–5 years)	0.00415	0.00384	0.00351	0.013	0.056	0.053
	(0.330)	(0.304)	(0.277)	(0.990)	(0.960)	(0.910)
Dummy = 1 if rural	0.153***	0.160***	0.166***	0.139***	0.57***	0.597***
	(6.649)	(6.902)	(7.083)	(9.290)	(10.510)	(10.760)
Constant	0.0668	0.0678	0.0689			
	(0.726)	(0.735)	(0.744)			
First-stage results						
Instrumental variable: ln(Euclidean distance)		0.791***			0.791***	
		(65.870)			(190.430)	
Instrumental variable: ln(natural path)			0.852***			0.851***
			(33.030)			(68.940)
Angrist-Pischke test of weak identification		4338.48	1091.24			
		P = 0.0000	P = 0.0000			
Observations	6,684	6,684	6,684	6,684	6,684	6,684

Sources: Nigeria Demographic and Health Survey (National Population Commission 2009); and calculations.
Note: Robust *t*-statistics clustered at the enumeration area in parentheses.
***$p < 0.01$, **$p < 0.05$, *$p < 0.10$.

Table 2B.8 Local GDP

Dependent variable: Local GDP	(1) Ordinary least squares, full sample	(2) Instrumental variable, natural path, full sample	(3) Instrumental variable, Euclidean distance, full sample	(4) Ordinary least squares, rural only	(5) Instrumental variable, natural path, rural only	(6) Instrumental variable, Euclidean distance, rural only
ln(transport cost to market)	-0.543***	-0.496***	-0.502***	-0.502***	-0.447***	-0.450***
	(-33.18)	(-26.28)	(-26.57)	(-29.03)	(-22.30)	(-22.41)
ln(distance to mine)	-0.0159	-0.0282	-0.0246	-0.0295	-0.0424*	-0.0395
	(-0.66)	(-1.17)	(-1.03)	(-1.20)	(-1.71)	(-1.60)
ln(population)	0.128***	0.101**	0.120***	0.00122	-0.00650	-0.00988
	(3.08)	(2.41)	(2.88)	(0.02)	(-0.13)	(-0.19)
ln(population)^2	0.0439***	0.0462***	0.0451***	0.0527***	0.0538***	0.0543***
	(17.79)	(18.50)	(18.21)	(16.24)	(16.46)	(16.70)
ln(cassava potential yield)	0.0191**	0.0160**	0.0176**	0.0191**	0.0155**	0.0172**
	(2.49)	(2.10)	(2.30)	(2.46)	(2.01)	(2.22)
ln(cassava potential yield)^2	0.00392**	0.00334*	0.00371**	0.00352*	0.00285	0.00326*
	(2.09)	(1.80)	(1.99)	(1.85)	(1.51)	(1.72)
ln(yams potential yield)	-0.0190**	-0.0175*	-0.0176*	-0.0190**	-0.0168*	-0.0172*
	(-2.07)	(-1.92)	(-1.93)	(-2.08)	(-1.84)	(-1.88)
ln(yams potential yield)^2	-0.00360	-0.00339	-0.00342	-0.00338	-0.00305	-0.00313
	(-1.60)	(-1.52)	(-1.53)	(-1.50)	(-1.36)	(-1.39)
ln(maize potential yield)	0.0639***	0.0603***	0.0649***	0.0749***	0.0699***	0.0762***
	(4.49)	(4.02)	(4.58)	(5.30)	(4.69)	(5.41)

(continued next page)

Table 2B.8 (continued)

Dependent variable: Local GDP	(1) Ordinary least squares, full sample	(2) Instrumental variable, natural path, full sample	(3) Instrumental variable, Euclidean distance, full sample	(4) Ordinary least squares, rural only	(5) Instrumental variable, natural path, rural only	(6) Instrumental variable, Euclidean distance, rural only
ln(maize potential yield)^2	−0.00717***	−0.00633***	−0.00693***	−0.00769***	−0.00668***	−0.00740***
	(−3.45)	(−2.94)	(−3.34)	(−3.73)	(−3.12)	(−3.59)
ln(rice potential yield)	−0.00606	−0.00638	−0.00620	−0.00661	−0.00700	−0.00692
	(−1.06)	(−1.12)	(−1.09)	(−1.14)	(−1.20)	(−1.19)
ln(rice potential yield)^2	0.000221	0.0000907	0.000218	0.000230	0.000105	0.000194
	(0.17)	(0.07)	(0.16)	(0.17)	(0.08)	(0.14)
Constant	−0.811***	−0.748***	−0.871***	0.397	0.397	0.334
	(−3.35)	(−3.09)	(−3.60)	(1.34)	(1.35)	(1.13)
Marketshed fixed effects	Yes	Yes	Yes	Yes	Yes	Yes
First-stage results						
ln(natural path)		0.7352***			0.7350***	
		(175.65)			(165.05)	
ln(Euclidean distance)			0.7330***			0.733***
			(176.29)			(165.60)
Angrist-Pischke test of weak identification		30,852.25	31,076.43		27,242.72	27,424.55
		P = 0.0000	P = 0.0000		P = 0.0000	P = 0.0000
Observations	10,728	10,607	10,728	9,899	9,797	9,899

Sources: Ghosh et al. 2010; LandScan 2006; FAO and IIASA 2000; and calculations.
Note: t-statistics in parentheses.
***p < 0.01, **p < 0.05, *p < 0.10.

Table 2B.9 Local GDP Spatial Bootstrapping

	(1)	(2)
Dependent variable: Local GDP	Instrumental variable, natural path, full sample	Instrumental variable, natural path, bootstrapped
ln(transport cost to market)	−0.496***	−0.462***
	(−26.28)	(−4.39)
ln(distance to mine)	−0.0282	−0.0254
	(−1.17)	(0.23)
ln(population)	0.101**	0.0226
	(2.41)	(−0.06)
ln(population)^2	0.0462***	0.0523***
	(18.50)	(−2.61)
ln(cassava potential yield)	0.0160**	0.0155
	(2.10)	(−0.39)
ln(cassava potential yield)^2	0.00334*	0.0033
	(1.80)	(−0.33)
ln(yams potential yield)	−0.0175*	−0.0168
	(−1.92)	(0.42)
ln(yams potential yield)^2	−0.00339	−0.0035
	(−1.52)	(0.35)
ln(maize potential yield)	0.0603***	0.1043
	(4.02)	(−0.31)
ln(maize potential yield)^2	−0.0063***	−0.0104
	(−2.94)	(0.35)
ln(rice potential yield)	−0.00638	−0.0056
	(−1.12)	(0.19)
ln(rice potential yield)^2	0.00009	0.0001
	(0.07)	(−0.01)
Constant	−0.748***	−0.8026
	(−3.09)	(0.38)
Marketshed fixed effects	Yes	Yes
Observations	10,607	Observations/sample: 500 Samples: 1,000

Sources: Ghosh et al. 2010; LandScan 2006; FAO and IISA 2000; and calculations.
Note: t-statistics in parentheses.
***$p < 0.01$, **$p < 0.05$, *$p < 0.10$.

Table 2B.10 Conley Bounds

	Support for possible values of δ	95 percent confidence interval	
	Instrumental variable: ln(natural path)	Lower bound	Upper bound
ln(crop revenue)	δ: [−0.0001, 0.0001]	−0.720	−0.123
	δ: [−0.001, 0.001]	−0.721	−0.122
	δ: [−0.01, 0.01]	−0.734	−0.108
ln(livestock sales)	δ: [−0.0001, 0.0001]	−0.624	0.055
	δ: [−0.001, 0.001]	−0.625	0.057
	δ: [−0.01, 0.01]	−0.638	0.070
ln(nonagricultural income)	δ: [−0.0001, 0.0001]	−0.650	−0.077
	δ: [−0.001, 0.001]	−0.651	−0.076
	δ: [−0.01, 0.01]	−0.662	−0.065
ln(wealth index)	δ: [−0.0001, 0.0001]	−0.238	−0.137
	δ: [−0.001, 0.001]	−0.239	−0.136
	δ: [−0.01, 0.01]	−0.251	−0.124
Multidimensional poverty index	δ: [−0.0001, 0.0001]	0.039	0.089
	δ: [−0.001, 0.001]	0.038	0.090
	δ: [−0.01, 0.01]	0.026	0.101
ln(local GDP)	δ: [−0.0001, 0.0001]	−0.533	−0.459
full sample	δ: [−0.001, 0.001]	−0.535	−0.458
	δ: [−0.01, 0.01]	−0.547	−0.445
ln(local GDP)	δ: [−0.0001, 0.0001]	−0.486	−0.407
rural only	δ: [−0.001, 0.001]	−0.488	−0.406
	δ: [−0.01, 0.01]	−0.500	−0.394
Year-round employment (male)	δ: [−0.0001, 0.0001]	−0.081	−0.015
	δ: [−0.001, 0.001]	−0.082	−0.014
	δ: [−0.01, 0.01]	−0.093	−0.003
Year-round employment (female)	δ: [−0.0001, 0.0001]	−0.093	−0.052
	δ: [−0.001, 0.001]	−0.094	−0.051
	δ: [−0.01, 0.01]	−0.105	−0.040
Agricultural employment (male)	δ: [−0.0001, 0.0001]	0.048	0.112
	δ: [−0.001, 0.001]	0.047	0.114
	δ: [−0.01, 0.01]	0.036	0.125
Agricultural employment (female)	δ: [−0.0001, 0.0001]	0.040	0.089
	δ: [−0.001, 0.001]	0.039	0.090
	δ: [−0.01, 0.01]	0.028	0.102

Source: World Bank calculations.
Note: Confidence intervals calculated following Conley, Hansen, and Rossi (2012).

Annex 2C Multidimensional Poverty Index

Table 2C.1 Multidimensional Poverty Index Components from Nigeria Demographic and Health Survey

Dimension	Indicator	Deprivation factor	Relative weight
Education	Highest degree earned	No household member has completed five years of education.	1.67
	Child school attendance	Household has a school-age child not attending school.	1.67
Health	Child mortality	Household has had at least one child age 0–5 years die in the past five years.	1.67
	Nutrition	Household has a malnourished woman age 15–49 or child age 0–5.	1.67
Standard of living	Electricity	The household has no electricity.	0.56
	Improved sanitation	Household does not have improved sanitation.	0.56
	Safe drinking water	Household does not have access to improved water source.	0.56
	Flooring	The household has a dirt floor.	0.56
	Cooking fuel	The household uses dirty cooking fuel.	0.56
	Asset ownership	The household does not own more than one bicycle, motorcycle, radio, refrigerator, TV, or phone, and does not own a car.	0.56

Note: World Health Organization standards were followed in determining what constitutes unimproved water sources, inadequate sanitation, and dirty cooking fuel.

A household is considered to be multidimensionally poor if its weighted sum of indicators is greater than 3. Note that the weights add up to about 10, which is the number of indicators (difference due to rounding).

Notes

1. The data are from DeLorme, a company specializing in constructing georeferenced data.
2. To "ground truth" and take advantage of firsthand local knowledge, government offices across Nigeria were surveyed about the conditions of specific road segments near them.
3. An alternative methodology in some research is to use case studies of natural experiments where road placement is "plausibly exogenous" (for example, where a road is built specifically for military purposes or to connect a mine with a port, with arguably no consideration for the area between the road's end-points) to see what happens to economic development in non-target regions along the road.
4. Technically, the condition is that A is not correlated with the error term from the regression.

5. The distance is likely to be correlated with the costs of building the road; the farther away two points are from each other, the longer the road must be to connect them, so the more expensive must be the construction. But there is no reason to believe that the costs and benefits are highly correlated.
6. This approach is similar to that of Faber's (2014) use of a hypothetical least-cost path.
7. The 2010 Living Standards Measurement Study–Integrated Surveys on Agriculture (LSMS-ISA) for Nigeria and the 2008 Nigeria Demographic and Health Survey (NDHS).
8. Light intensity has been found to be highly correlated with GDP (Ghosh et al. 2010).
9. Measured flow benefits for long-established roads would reflect the long-term equilibrium, but the data do not allow "old" roads to be distinguished from "new."
10. See annex 2C for details on how the MPI was calculated.
11. Given the interrelatedness of the three sources of income (crop revenue, livestock sales, and nonagricultural income), a seemingly unrelated regression and a three-stage least squares model were estimated. These estimates are broadly consistent.
12. The bootstrapping technique involves first determining the bandwidth of the short-distance residual autocorrelation (RSA) from the local GDP regressions. The data are then resampled 1,000 times in a way that ensures that no two points in the same sample are within that RSA bandwidth of each other, ensuring that each sample is spatially independent. Bootstrapped parameter estimates are then generated from these samples, and the estimates they provide are compared with the full sample estimates. Because no statistical distinction can be made between the full sample estimates and the spatial independent bootstrapped estimate, it is determined that there is no bias due to spatial autocorrelation in the full sample estimates. See table 2B.9 in annex 2B for results, and Ali et al. (2015) for a detailed explanation of the spatial bootstrapping process.
13. This report analyzes the combined effect of both large transport infrastructure, such as highways, and rural roads.
14. There are six geopolitical zones in Nigeria: South South, South West, South East, North Central, North East, and North West. These are political divisions that were formed in the 1960s along tribal lines. Using these political lines is one way to allow for the heterogeneous spatial effects of transport.
15. Formally, local GDP increase is calculated as follows:

$$B_j = \sum_i \eta_k \times \tau_{ij} \times y_i$$

in which B_j is the total increase in local GDP due to project j, η_k is the local GDP elasticity of transport costs for region k (from table 2.2), τ_{ij} is the percentage change in transport costs in cell i due to project j, and y_i is the baseline GDP in cell i from the local GDP data.
16. Road improvement costs per kilometer will likely not be uniform, but this assumption was made merely for demonstration purposes. If road improvement costs are known or estimated, a benefit-cost ratio could easily be calculated.

References

Ali, Rubaba, Federico Barra, Claudia Berg, Richard Damania, John Nash, and Jason Russ. 2015. "Transport Infrastructure and Welfare: An Application to Nigeria." Policy Research Working Paper 7271, World Bank, Washington, DC.

Alkire, Sabina, and Maria Emma Santos. 2010. "Acute Multidimensional Poverty: A New Index for Developing Countries." OPHI Working Paper 38, University of Oxford.

Conley, Timothy G., Christian B. Hansen, and Peter E. Rossi. 2012. "Plausibly Exogenous." *Review of Economics and Statistics* 94 (1): 260–72.

Dorosh, Paul, Hyoung Gun Wang, Liangzhi You, and Emily Schmidt. 2012. "Road Connectivity, Population, and Crop Production in Sub-Saharan Africa." *Agricultural Economics* 43 (1): 89–103.

Emran, Shahe, Fenohasina Maret-Rakotondrazaka, and Stephen Smith. 2014. "Education and Freedom of Choice: Evidence from Arranged Marriages in Vietnam." *Journal of Development Studies* 50 (4): 481–501.

Faber, Benjamin. 2014. "Trade Integration, Market Size, and Industrialization: Evidence from China's National Trunk Highway System." *Review of Economic Studies* 81: 1046–70.

FAO and IIASA (Food and Agriculture Organization of the United Nations and the International Institute for Applied Systems Analysis). 2000. "Global Agro-Ecological Zones (GAEZ)." FAO, Rome; and IIASA, Laxenberg, Austria.

Ghosh, T., R. L. Powell, C. D. Elvidge, K. E. Baugh, P. C. Sutton, and S. Anderson. 2010. "Shedding Light on the Global Distribution of Economic Activity." *Open Geography Journal* 3 (1): 148–61.

Landscan. 2006. Global Population Database (2006 release). Oak Ridge National Laboratory, Oak Ridge, Tennessee. http://www.ornl.gov/landscan/.

National Bureau of Statistics, Federal Republic of Nigeria. Nigeria General Household Survey (GHS), Panel 2010, Ref. NGA_2010_GHS_v02_M. Dataset downloaded from http://econ.worldbank.org/WBSITE/EXTERNAL/EXTDEC/EXTRESEARCH/EXTLSMS/0,,contentMDK:22949589~menuPK:4196952~pagePK:64168445~piPK:64168309~theSitePK:3358997,00.html.

National Population Commission (NPC) [Nigeria] and ICF Macro. 2009. "Nigeria Demographic and Health Survey 2008." National Population Commission and ICF Macro, Abuja, Nigeria.

Ulimwengu, John, Jose Funes, Derek D. Headey, and Liang You. 2009. "Paving the Way for Development: The Impact of Road Infrastructure on Agricultural Production and Household Wealth in the Democratic Republic of Congo." Agricultural and Applied Economics Association Annual Meeting, Milwaukee, Wisconsin, July 26–28.

World Bank. 2013. "Unlocking Africa's Agricultural Potential: An Action Agenda for Transformation." Africa Region Sustainable Development Series. World Bank, Washington, DC.

Chapter 3

Impact of Transport Cost on Technology Adoption

Introduction

This chapter seeks to address an old and recurring theme in development economics—that of the slow adoption of new technologies by farmers in many developing countries despite the potential for higher returns. The chapter extends the analysis in chapter 2 by exploring an underlying mechanism through which lower transport costs could influence the choice of production technology and the differential impact on returns of modern versus traditional farming techniques. It demonstrates that the constraints to the adoption of modern technologies and access to markets are linked and should as a consequence be targeted jointly.

Motivation

Agricultural yields across many parts of Sub-Saharan Africa have stagnated and even declined in some countries, despite the availability of new agricultural technologies, improvements in transportation infrastructure, and policy reforms aimed at market liberalization. A comparison of Africa's performance with that in South America and three subregions of Asia over the course of two decades indicates that Africa has exhibited lower total factor productivity growth overall than its counterparts in the 1990s, then fell even further behind in the 2000s, magnifying the total factor productivity gap (World Bank [2013], based on Fuglie, Wang, and Ball [2012]). As an example, cereal yields in Sub-Saharan Africa grew at an average annual rate of 0.7 percent between 1980 and 2000, compared with other developing regions that saw growth ranging from 1.2 to 2.3 percent (Slootmaker 2013). Similarly, the average yield of tuber crops in Sub-Saharan Africa is the lowest in the world at about 8 tons per hectare (Pinstrup-Andersen, Pandya-Lorch, and Rosegrant 1997). Since the 1980s fertilizer intensity (kilograms per hectare of farmland) in Sub-Saharan

Africa grew, on average, by 0.93 percent annually compared with 5 percent in South Asia (Slootmaker 2013).[1] Furthermore, Gollin, Morris, and Byerlee (2005) note that by 2000, modern maize varieties only accounted for an estimated 17 percent of total maize production in Africa, compared with 90 percent in East and Southeast Asia and 57 percent in Latin America. With a population that is set to double by 2040 at current trends, these are worrying signs for a region that aspires to generate food surpluses.

Numerous explanations have been offered for the low productivity of agriculture in Africa. Some work has focused on the need for price reforms to sharpen producer incentives and promote greater competition and efficiency (World Bank 2012). Other studies highlight the role of transaction costs, which, if reduced, would raise farm-level prices. If there are impediments to market participation, price incentives will be dampened and may even be rendered ineffectual (Schiff and Montenegro 1997).[2] Another strand of literature studies the reasons for the slow adoption of modern technologies and advances numerous explanations, including learning impediments, credit constraints, risk, and differences among farmers, or views low fertilizer use in Africa as a rational decision due to high fertilizer prices combined with low response rates in some African soils.[3] Many of these explanations are not mutually exclusive, and it is plausible that several play a role.

Theory and Hypotheses

This chapter offers a somewhat novel explanation for the slow adoption of improved technology in many developing countries. To guide the empirical analysis, a theoretical model is developed to explore the possible links between access to markets and technology choice. Box 3.1 provides a short description of the theoretical model, and annex 3A the full version. It is assumed that there are nonconvexities such as fixed costs or minimum costs to adopting more modern inputs. Examples include the fixed costs of, say, drilling a well for irrigation, or the minimum price that is usually paid in machinery rental, or information and learning costs associated with learning how to use a new technology or grow a new crop. These fixed costs and nonconvexities create a hurdle that households must overcome to adopt the more productive technology. Transport costs affect the returns to technology adoption and hence the likelihood that this hurdle will be crossed. Hence, variations in transport costs generate differences in payoffs and influence the choice of technology.

The theoretical analysis suggests two testable hypotheses: (1) that lowering transport costs will induce higher adoption of modern technology and (2) that agricultural income is more responsive to reductions in transport costs for farmers already employing modern farming technologies. The rationale behind

Theoretical Model

There are two types of farmers in the model, traditional farmers who use a less productive technology and those who use an improved technology such as better (or any) machinery. Use of the improved technology requires payment of a fixed cost (*F*), which enhances the productivity of farming. *F* could be interpreted broadly to represent a variety of impediments to adoption—a threshold price on the rental of machinery, learning costs, technological lumpiness, and so on. In all other respects the farmers are identical. There is only one period with two stages. In the first stage each farmer independently decides whether to pay the fixed costs and adopt the productivity-enhancing technology or remain with the traditional technology, and in the second stage each farmer determines how much to produce and how much of this output to sell in a market (or conversely to consume domestically).

Upon solving the model via backward induction, three main insights can be drawn. First, the model suggests that adoption of new technologies will be more pervasive where transport costs are lower (Result 1). Next, the model predicts that reductions in transport costs will have a larger impact on the marketed output of farmers using modern farming techniques (Result 2). Finally, farmers with more modern technologies are likely to be better integrated into markets. See annex 3A for the full derivation of this model.

the first prediction is self-evident. Lower transport costs raise profits and, all else equal, render recovering the fixed costs of technology adoption more likely, thus inducing a switch from less to more modern forms of agriculture. The second prediction reflects the fact that farmers using more modern techniques produce a greater surplus and sell a larger amount of their output in markets. As a result their exposure and responsiveness to changes in market conditions is greater. Traditional farmers, in contrast, tend to engage in subsistence production and sell a smaller fraction of their crops, if any, in markets. As a consequence they are less affected by variations in transport cost. This theoretical framework fits into a larger family of models of technology choices developed by Mundlak (1988) and later applied by Mundlak, Larson, and Butzer (1999), Mundlak, Butzer, and Larson (2012), and Larson and León (2006). Essentially, farmers face different circumstances (different roads, markets, climates, and so forth) and so choose different technologies to maximize their profits. By making technology choices, farmers move between production functions as well as along them.

To test these hypotheses, two forms of data are used. The first data set, from the Spatial Production Allocation Model (SPAM; HarvestChoice 2012), contains a cross-section of gridded agricultural production data for several crops. This data set has the benefit of clearly distinguishing between crops produced using traditional methods and crops produced using modern techniques. However, the SPAM data, although useful, are limited since they are spatially aggregated and cannot be used to test whether variations in transport costs induce changes in technology choice at the farm level. The second data set, from the Living Standards Measurement Study—Integrated Surveys on Agriculture (LSMS-ISA), is a household survey that contains specific information about agricultural producers. This data set allows for a deeper analysis of the interaction between modern farming techniques and transport costs, but does not clearly define the choice of production technology, though it does provide information on input usage that can be used as a proxy for technology decisions. In the analysis that follows, machinery use is used as a measure of technology choice for reasons that are explained in greater detail in the following section.

Estimating Impacts of Roads on Technology Adoption Using SPAM Data

To motivate the empirical analysis, spatial data are used to determine the impact of transport costs on crop production under different input systems. Data from the SPAM are particularly useful for the purposes herein given that the model clearly distinguishes between the production of various crops under different input systems.[4] These categories map into conventional definitions of traditional and modern production. Examining the effects of transport costs on traditional and modern input production systems separately provides a first, albeit incomplete, test of the possibly differential impacts of transport costs on agricultural output.

An agricultural production function is estimated for four crops that account for more than 60 percent of the total agricultural production value in Nigeria in 2011, according to the Food and Agriculture Organization (FAO) of the United Nations:[5] yams, rice, cassava, and maize (see box 3.2 for details on the SPAM estimation strategy). Cassava and maize are typically grown for home consumption using traditional low-input systems (in fact, in Nigeria, neither of these crops is produced using high-input or irrigated technologies, according to SPAM), whereas yams and rice are more commonly marketed and are grown under both traditional low-input systems and high-input management regimes.

The first, and perhaps most intuitive, result is that reducing transport costs leads to greater *high-input* production of yams and rice (shown in columns 1 and 2 of table 3.1, respectively). To be precise, for each 10 percent reduction

SPAM Model

The data used in this analysis are spatially organized into a gridded framework. The total land area of Nigeria is split into pixels of 5 arcminute × 5 arcminute (approximately 10 kilometers × 10 kilometers), which line up with the SPAM cells described in the main text. Each pixel is a unique observation. An agricultural production function is estimated for four crops—yams, rice, cassava, and maize—and takes the following form:

$$\ln\left(Y_{ik}^{j}\right) = \beta_{0k}^{j} + \beta_{1k}^{j}\ln\left(T_i\right) + X_{ik}^{j\prime}\delta^{j} + \varepsilon_{ik}^{j}, \tag{1}$$

in which Y_{ik}^{j} is total production of crop j in pixel i using input system k, and k can be traditional (subsistence, as coded by SPAM) or modern (irrigated) inputs. T_i is transport costs from cell i to the closest market, with a market being defined as a city of at least 100,000 residents, and X_{ik}^{j} is a vector of control variables, which are discussed below. By measuring the cost to the nearest market, rather than distance or travel time as is more common in the literature, farmers' decisions about where they will transport their goods can be more accurately determined. To correct for potential bias due to the non-random placement of roads, equation (1) is estimated using two-stage least squares, where the natural path variable, described in appendix A on Geospatial Analysis, is used as the instrument for transport costs.

Several spatial data sets were used or generated to control for confounding variables, including population, agro-ecological production potential (yield), and distance to mining facilities. Population data come from LandScan (2006),[a] which uses satellite imagery analysis to disaggregate census data into a gridded network. Agro-ecological potential data are from Global Agro-Ecological Zones (FAO and IIASA 2000), which considers climate and soil conditions to estimate the maximum potential yield in each pixel, for each crop.[b] Euclidean distance to the nearest mining facility is calculated using data from the National Minerals Information Center of the United States Geological Survey, and is included to account for the fact that mining facilities often have high concentrations of workers and their families, making them very high demand centers.[c] Because this analysis is focused on agricultural production, which mainly occurs in rural areas, urban areas were removed from the data set.[d] Additionally, cells in which the production potential (according to Global Agro-Ecological Zones) of the crop being estimated is zero are omitted. Cells with a positive production potential, but zero actual production, are left in the model and treated as true zeros.

(continued next page)

Box 3.2 (continued)

See annex 3B, table 3B.5 for summary statistics.

a. LandScan population data are available here: http://web.ornl.gov/sci/landscan/landscan _data_avail.shtml.

b. The data used in this model assume climatic conditions similar to the 1961–90 baseline level. Global Agro-Ecological Zones also differentiates potential yields by input system. For the traditional-input equation, potential yields under low inputs were used, and for the modern-input equation, potential yields under high inputs were used.

c. Only a subset of mining facilities available in the raw data are considered. Facilities selected were those that involve the extraction of minerals or hydrocarbons from the ground (specifically coal, tin, iron, nitrogen, and petroleum) or the processing of hydrocarbons. Mining facilities that were in the United States Geological Survey data set but not included in this analysis include facilities, such as cement plants or steel mills, that are likely concentrated in large cities or manufacturing areas. In addition, only plants that were active between 2006 and 2010 were included.

d. The methodology for determining which pixels of the LandScan data set are urban areas is as follows: Nigeria's urbanization rate as defined by the Central Intelligence Agency World Factbook was 49.6 percent in 2011 (https://www.cia.gov/library/publications/the-world -factbook/fields/2212.html). The total population in the LandScan data set is approximately 136 million, implying an urban population of 67 million. The pixels with the largest number of people according to LandScan are marked as being urban pixels until the total number of people living in these marked pixels equals 67 million. These marked pixels are then omitted from the regressions. Aside from the above exceptions, and pixels falling entirely in water, all other pixels within Nigeria's borders are included in the analysis.

Table 3.1 SPAM High-Input Production

Two-stage least squares	(1) Dependent variable: ln(yams high-input production)	(2) Dependent variable: ln(rice high-input production)
ln(transport cost to market)	−0.310***	−0.528***
	(−2.81)	(−3.96)
Other controls (all in log form)	Population, population^2, distance to mine, high-input potential yield, high-input potential yield^2, state fixed effects	
Instrumental variable	Natural path	Natural path
Observations	8,783	6,449

Sources: SPAM (HarvestChoice 2012); LandScan 2006; FAO and IIASA 2000; and calculations.
Note: SPAM = Spatial Production Allocation Model. For full results, see annex 3B, tables 3B.1 and 3B.2.
t-statistics in parentheses.
***$p < 0.01$, **$p < 0.05$, *$p < 0.10$.

in transport costs, all else equal, high-input yam production increases by 3.1 percent and high-input rice production increases by 5.3 percent. Conversely, *low-input* yam production does not change when transport costs are reduced, while low-input rice production actually declines by 5.2 percent for each 10 percent reduction in transport costs (shown in columns 1 and 2 of table 3.2, respectively).

Similar responses to transport costs are observed for cassava and maize. Low-input cassava output does not change when transport costs are reduced, whereas low-input production of maize decreases when transport costs decline—a 10 percent reduction in transport costs leads to a 12 percent reduction in low-input maize production[6] (shown in columns 3 and 4 of table 3.2, respectively).

In sum, reducing transport costs increases production of high-input crops, but has no effect, or possibly a negative effect, on low-input production. There are two plausible reasons why this relationship might hold. The first reason is that low-input (subsistence) farmers do not generate sufficient marketable surplus to benefit from cheaper transport to markets (consistent with Result 2 from box 3.1). This is a plausible explanation for the results observed for cassava and yams. The results for rice and maize, however, require a different explanation—not only does reducing transport costs not lead to increased low-input production of these two crops, it actually leads to decreased levels. A likely explanation is that when transport costs decline, farmers switch to more modern inputs (consistent with Result 1 from box 3.1). This leads to a reduction in low-input production in the areas under consideration, but aggregate

Table 3.2 SPAM Low-Input Production

Two-stage least squares	(1) Dependent variable: ln(yams low-input production)	(2) Dependent variable: ln(rice low-input production)	(3) Dependent variable: ln(cassava low-input production)	(4) Dependent variable: ln(maize low-input production)
ln(transport cost to market)	0.0849	0.522***	−0.243	1.212***
	(0.64)	(3.79)	(−1.29)	(7.31)
Other controls (all in log form)	Population, population^2, distance to mine, low-input potential yield, low-input potential yield^2, state fixed effects			
Instrumental variable	Natural path	Natural path	Natural path	Natural path
Observations	8,649	6,679	7,435	9,144

Sources: SPAM (HarvestChoice 2012); LandScan 2006; FAO and IIASA 2000; and calculations.
Note: SPAM = Spatial Production Allocation Model. For full results, see annex 3B, tables 3B.1, 3B.2, 3B.3, and 3B.4.
t-statistics in parentheses.
***$p < 0.01$, **$p < 0.05$, *$p < 0.10$.

agricultural production likely actually increases—some production is simply shifted from low-input to high-input agriculture.[7]

The SPAM results presented here provide preliminary evidence of the relationship between agricultural production, input systems, and transport costs. However, because of the spatially aggregated nature of the data, directly assessing whether variations in transport costs induce shifts in technology choice from traditional to modern is not possible. For this reason, the next section analyzes household survey data from the LSMS-ISA. This rich household data set contains information on household agricultural revenue and use of mechanized inputs, among others, allowing for more direct tests of the impacts of transport costs on technology and production choices.

Estimating Impacts of Roads on Technology Adoption Using LSMS Data

Unlike with the SPAM analysis, household survey data do not clearly distinguish between low-input and high-input production. There are several reasonable dimensions along which technology choice can be defined; common candidates might include inputs such as fertilizers, high-yield seed varieties, pesticides and herbicides, and management regimes and the use of capital equipment such as irrigation or machinery. In Nigeria, as well as in many other developing countries, high-yield seed varieties, fertilizers, and other direct agricultural inputs are provided to farmers at heavily subsidized prices. Government policies therefore distort adoption decisions, making such inputs poor candidates for making the distinction between low-input and high-input production. Irrigation is also not a suitable indicator because it is more a function of where a farm is located (that is, proximity to a river or lake, average rainfall, and the like) rather than a decision variable. Machinery (such as tractors, plows, harvesters, carts, and so on), in contrast, must either be bought or rented and represents a significant investment, which is not typically subsidized. For the purposes of the LSMS analysis, modern farmers are defined as those who use mechanized inputs and traditional farmers are those who do not.

Table 3.3 reports the estimated impact of transport cost to market on the probability of modern technology adoption as proxied by machine use (see box 3.3 for details on the estimation process). A two-step procedure is used to explore responses to changes in transport costs. First, a model is estimated to capture the discrete choice of whether to adopt modern techniques (machinery use in this case). The endogeneity of transport costs is addressed in both equations by instrumenting with the natural path variable. Two specifications of the model are estimated by instrumental variable–probit in the interest of thoroughness.

Table 3.3 Machinery Use and Transport Cost to Market

Instrumental variable–probit Dependent variable Dummy = 1 if uses machinery	(1) Machine use dummy	(2) Machine use dummy
ln(transport cost to market)	−0.18***	−0.25***
	(−2.33)	(−3.28)
ln(nonagricultural income)	0.04**	
	(2.33)	
ln(neighborhood effects)		4.814***
		(3.86)
Other controls	Age, age², land, fertilizer use, distance to mine	
Instrumental variable	Natural path	Natural path
Observations	1,354	1,354

Sources: Nigeria Living Standards Measurement Study–Integrated Surveys on Agriculture (National Bureau of Statistics 2010); and calculations.
Note: For full results, see annex 3C, table 3C.1. Robust *t*-statistics in parentheses.
***$p < 0.01$, **$p < 0.05$, *$p < 0.10$.

LSMS Estimation Technique

A two-step estimation technique is followed to estimate the impact of reduced transport costs on (1) adoption of modern technology and (2) the agricultural income of modern versus traditional farmers.

To determine the impact of transport cost on adoption of modern technology, the following model is estimated:

$$dM_i = 1(\alpha_0 + \alpha_1 \ln(T_i) + \alpha_2 \ln(Z_i) + X_i'\delta + v_i > 0),$$ (1)

in which dM_i is a dummy taking the value of 1 if household i uses machinery and taking the value of zero if it does not, transport cost to market is given by T_i, Z_i is the exclusion restriction, and household characteristics are given by X_i. To account for potential treatment effects (described in box 3.4), two alternative exclusion restrictions are tested: nonagricultural income and average machinery use within the village. The parameter of interest is α_1, which indicates the causal effect of the cost of traveling to the closest market on the probability of using machinery. Equation (1) is estimated by instrumental variable–probit, instrumenting for transport costs with the natural path.

(continued next page)

Box 3.3 (continued)

Next, to determine whether transport costs have a differential impact on the revenues of modern versus traditional farmers, the following model is estimated:

$$\ln(R_i) = \beta_0 + \beta_1 \ln(T_i) + X_i'\gamma + \rho\lambda_i + u_i, \text{ if } dM_i = \{0,1\}, \qquad (2)$$

in which R_1 represents the household's agricultural income earned from crop sales, λ_i is the inverse Mills ratio (see box 3.4), and the remaining variables are the same as above. Equation (2) is estimated by two-stage least squares and, as before, the instrument for transport costs is the natural path variable.

Summary statistics for variables used in the estimation of these equations are given in annex 3C, table 3C.4.

Consider first the discrete choice. Column 1 of table 3.3 presents the instrumental variable–probit results from estimating the effect of transport cost on the probability of modern technology using the household's nonagricultural income as the instrumental variable. There are typically indivisibilities in the purchase or rental of machinery (such as a minimum cost). Therefore, farmers with greater endowments of nonagricultural income would be better positioned to pay these costs, suggesting that nonagricultural income should positively influence the likelihood of machinery adoption. Column 1 presents estimates using nonagricultural income as the exclusion restriction (as explained in box 3.4).

Column 2 of table 3.3 presents the instrumental variable–probit results using a different exclusion restriction in the interests of robustness. It is well established in the literature that neighborhood effects strongly influence a household's adoption of new technologies. For example, Conley and Udry (2010) find strong evidence that pineapple growers in Ghana adjust their fertilizer use on the basis of their neighbors' experiences. Thus, neighborhood effects are expected to positively influence the likelihood of adopting modern technology. Average machinery use within the village is the second exclusion restriction used in the model.[8]

The estimates from the two specifications are quite similar and are consistent with the theoretical model: lowering transport costs by 10 percent increases the probability that a farmer uses mechanized farming techniques by 1.8–2.5 percent. This is persuasive evidence that farmers switch from non-mechanized to mechanized farming as transport costs are reduced.

Estimating the Differential Impact of Roads on Modern versus Traditional Farmers

Next, the differential impacts of roads on modern versus traditional farmers are explored. This second stage involves a continuous choice of how much to produce, contingent upon technology choice, transport costs, and other exogenous variables. Examining the subsamples of modern and traditional farmers

Treatment Effects Model

When estimating the differential treatment effect on two subgroups, that is, mechanized vs. nonmechanized farmers in this analysis, heterogeneous treatment effects must be accounted for to prevent estimates from being biased. This is done by estimating a treatment effects model, whereby the inverse Mills ratio (λ_v) is included in the second stage, representing the probability of being included in the sample, following the Heckman two-step estimation process (Heckman 1976). The inverse Mills ratio is calculated from the fitted values from estimating equation (1) in box 3.3 by probit (after proxying for transport costs with the natural path).

Two alternate exclusion restrictions are employed: nonagricultural income and neighborhood effects (average machine use within the village). Nonagricultural income and neighborhood effects are expected to increase the probability of a farmer's machine use (equation (1) in box 3.3), but not to have any direct influence on that farmer's revenue. Intuitively, greater nonagricultural income should facilitate payment of the fixed costs of technology adoption. After taking into account other inputs that could be bought to increase crop revenue, for example, fertilizer, the income should not directly influence crop revenue. In the case of neighborhood effects, use of a tractor by a farmer's neighbor might make it more likely that (s)he would also use one, but should not directly influence revenue.

Table B3.4.1 reports the estimated impact of transport cost to market on the crop revenue of modern farmers. Column 1 includes inverse Mills ratio 1 (IMR_1), which was calculated using nonagricultural income, and column 2 includes IMR_2, which was computed using machinery-use neighborhood effects. The estimated impact of transport cost to market is similar in both cases. A 10 percent reduction in transport costs would increase crop revenue

(continued next page)

Box 3.4 (continued)

by 22 and 25 percent, respectively. Note that the two inverse Mills ratios (IMR_1 and IMR_2) are not significant, which suggests that selection bias is not present, that is, there are no omitted variables that influence selection into mechanization. Thus, the impact of transport costs on the two groups separately can be estimated separately, as is done in the table.

Table B3.4.1 Heckman Two-Step

Two-stage least squares	(1) Dependent variable: ln(crop revenue) Modern farmers	(2) Dependent variable: ln(crop revenue) Modern farmers
ln(transport cost to market)	−2.20***	−2.46***
	(−3.43)	(−6.54)
IMR_1 (nonagricultural income)	−0.413	
	(−0.12)	
IMR_2 (neighborhood effects)		−0.193
		(−0.56)
Other controls	Age, age², land, labor, labor², fertilizer use, distance to mine, irrigation, and agro-ecological zone fixed effect	
Instrumental variable	Natural path	Natural path
Observations	393	621

Sources: Nigeria Living Standards Measurement Study–Integrated Surveys on Agriculture (National Bureau of Statistics 2010); and calculations.
Note: IMR= inverse Mills ratio (see Heckman 1976). For full results, see annex 3C, Table 3C.2. Robust *t*-statistics in parentheses.
***$p < 0.01$, **$p < 0.05$, *$p < 0.10$.

separately could potentially induce selection bias. This is tested for and rejected using the Heckman procedure described in box 3.4. Table 3.4 reports the instrumental variable estimates of the effect of transport costs on the crop revenue of the different subsamples of farmers (mechanized and nonmechanized) and compares them with the sample as a whole. Column 1 estimates the model for the full sample, column 2 focuses on the subsample of mechanized farmers, and column 3 focuses on traditional farmers. As predicted by theory, the coefficient on transport costs is significantly lower in absolute value for nonmechanized farmers (1.4) than it is for mechanized farmers (2.4). This result is evidence that reducing transport costs benefits farmers using modern inputs significantly

Table 3.4 Differential Impact of Roads on Modern versus Traditional Farmers

	(1)	(2)	(3)
	Dependent variable: ln(crop revenue) All farmers	Dependent variable: ln(crop revenue) Modern farmers	Dependent variable: ln(crop revenue) Traditional farmers
ln(transport cost to market)	−1.77***	−2.44***	−1.42***
	(−8.44)	(−6.57)	(−5.47)
Other controls	Age, age², land, labor, labor², fertilizer use, distance to mine, irrigation, and agro-ecological zone fixed effects		
Instrumental variable	Natural path	Natural path	Natural path
Observations	2,598	624	1,974

Sources: Nigeria Living Standards Measurement Study–Integrated Surveys on Agriculture (National Bureau of Statistics 2010); and calculations.
Note: For full results, see annex 3C, table 3C.3. Robust t-statistics in parentheses.
***$p < 0.01$, **$p < 0.05$, *$p < 0.10$.

more than traditional-input farmers. (See annex 3C for further details relating to the LSMS regression results.)

In sum, the empirical assessment provides compelling evidence to support the hypotheses that reducing transport costs will encourage farmers to switch to more mechanized forms of agriculture, and that the mechanized farmers are more sensitive to variations in transport costs. These findings are robust to alternative specifications and to controlling for sources of endogeneity and selection biases.

Conclusions

Although it has long been recognized in the literature that reduction of transport costs will induce greater market participation and increase welfare, this chapter takes a fresh look at the underlying mechanisms. A hitherto neglected channel through which improved access may induce the adoption of more modern technologies is identified.

Two sources of data—SPAM and LSMS-ISA—are used to assess the impact of transport costs on the adoption of modern inputs as well as the potentially differential effects on modern versus traditional farmers. The endogeneity of transport costs is addressed using a novel instrumental variable, the natural path, which is arguably an improvement over the straight-line instrumental variables commonly used in the literature (see discussion in chapter 2, box 2.1). Robust evidence is presented that a reduction of transport costs will increase the adoption of modern technologies. Furthermore, the results indicate that transport costs have a greater impact on modern than on traditional farmers.

The SPAM analysis reveals that when transport costs are reduced, crop production under a high-input regime is increased, whereas crops produced under a low-input regime are either unaffected or may experience a decline in output. It is likely that this decline in low-input production is explained by the switch toward more modern agriculture. The LSMS analysis, in turn, yields evidence supporting the predictions from theory (box 3.1) that reducing transport costs increases the likelihood of modern technology use. Further evidence demonstrates that all farmers (both modern and traditional) see an increase in crop revenue when transport costs decline, but modern farmers' revenue increases by a significantly greater margin.

In sum, this chapter presents compelling evidence that the constraints to the adoption of modern technologies and access to markets are interconnected and therefore should be targeted jointly. For example, reducing transport costs to the market may not be sufficient to push the local economy toward a more favorable equilibrium. It may be necessary to also expand access to credit, which would enable farmers to cover the fixed costs involved in modernizing. By the same token, expanding credit by itself may not be enough of a push either, if market access is insufficient or too costly. To increase yields in Sub-Saharan Africa, policies that bundle these interventions, focusing on improving both technology availability and connectivity, are needed.

Annex 3A Theoretical Model

This annex outlines a minimalist model that describes the manner in which responses might diverge between different types of farmers. The model distinguishes between traditional farmers who use a lower-productivity technology and those who adopt improved technology that generates higher payoffs and yields. In contrast to the existing literature, the focus is not on differences in factor endowments, risk, or imperfections in capital or labor markets that might lead to differential responses. Instead, the model assumes identical endowments, and differences emerge simply from nonconvexities—specifically the need to cover the fixed costs of accessing a more productive technology. Such nonconvexities may emerge either from the technology itself (minimum size or other indivisibility), nonlinear pricing (when a machine is rented for a minimum amount of time), or information and learning costs (learning how to drive a tractor or grow a new crop), and other forms of inertia.[9] Analogous to the big-push literature,[10] a switch from traditional to more modern farming occurs if the payoffs from switching exceed the fixed costs of adopting the new technology. Unlike the big-push models the outcome does not necessarily depend on the decisions of other farmers, demand factors, or market failures. This theoretical framework fits into a larger family of models of technology choices developed by Mundlak (1988) and later applied by Mundlak, Larson, and Butzer (1999), Mundlak, Butzer, and Larson (2012), and Larson and León (2006). Essentially, farmers face different circumstances (different roads, markets, climates, and so on) and therefore select different technologies to maximize their profit. By making technology choices, farmers move between production functions as well as along them.

For brevity a stylized version of the model that underlies the empirical work is outlined, based on specific functional forms that yield closed-form solutions. Generalizations without these functional forms are straightforward.

There are two types of farmers in the model, traditional farmers who use a less productive technology and those with an improved technology such as access to better (or any) machinery. Use of the improved technology requires payment of a fixed cost (F) that enhances the productivity of farming. As noted earlier, F could be interpreted broadly to represent a variety of impediments to adoption—a threshold price on the rental of machinery, learning costs, technological lumpiness, and so on. In all other respects the farmers are identical. The fixed costs may be financial or nonpecuniary, and may arise for a number of reasons that have been frequently documented in the literature.

There is only one period with two stages. Production decisions are made sequentially. In the first stage each farmer independently decides whether to pay the fixed costs and adopt the productivity-enhancing technology or remain with the traditional technology. Having made this technology decision, in the second stage each farmer determines how much to produce and how much of this output to sell in a market (or conversely to consume domestically). For simplicity, the model starts with the case of two farmers indexed i (more modern) and j (traditional). By backward induction the final stage is solved first.

Modern Farmers

The utility function of the modern farmer is simply given by equation (3.1):

$$U_i = B_i^\beta (\delta_i y_i)^\alpha \tag{3.1}$$

in which B_i is the quantity of goods bought from the market for consumption by farmer i, δ_i is the proportion of goods produced, y_i, that are consumed domestically (that is, the amount sold to markets is $(1 - \delta_i)y_i$). Equation (3.1) is maximized subject to the following constraints: the production function is

$y_i = W_i^\gamma$, where $1 > \gamma > 0$ and the input W is supplied under competitive markets at price v. The budget constraint is expressed as in equation (3.2):

$$(1-\delta_i)(p-t)W_i^\gamma = (p_B + t_B)B_i + vW + F, \qquad (3.2)$$

in which t and t_B are transport costs for goods y and B, respectively, and p and p_B are the given market prices of farm output and consumption goods, respectively. F is the fixed cost for using technology $y_i = W_i^\gamma$, and v is the variable cost. By equation (3.2) sales of goods produced $\left((1-\delta_i)(p-t)W_i^\gamma\right)$ must equal total expenditures, that is, the sum of spending for consumption goods $\left((p_B + t_B)B_i\right)$, the purchase of the input (vW), and the fixed cost for using the improved technology (F). Maximizing (3.1) subject to (3.2) yields the following first-order-conditions:

$$\frac{dL}{dB} = \beta B^{\beta-1}\delta_i^\alpha W_i^\varepsilon - \lambda(p_B + t_B) = 0$$

in which $\varepsilon = \alpha\gamma$

$$\frac{dL}{dW} = \varepsilon B^\beta \delta_i^\alpha W_i^{\varepsilon-1} + \lambda\left(\varepsilon(p-t)(1-\delta_i)W_i^{\gamma-1} - v\right) = 0$$

$$\frac{dL}{d\delta_i} = \alpha B^\beta \delta_i^{\alpha-1} W_i^\varepsilon - \lambda\left((p-t)W_i^\gamma\right) = 0$$

$$\frac{dL}{d\lambda} = (1-\delta_i)(p-t)W_i^\gamma - (p_B + t_B)B_i - vW - F = 0.$$

These equations can be solved for the endogenous variables:

$$W_i = \left(\frac{\gamma(p-t)}{v}\right)^{\frac{1}{1-\gamma}}$$

$$\delta_i = \frac{\varepsilon(p_B - t_B)}{\beta}\left(\frac{1}{(p-t)}\left(\frac{v}{\gamma}\right)^\gamma\right)^{\frac{1}{1-\gamma}} \qquad (3.3)$$

$$B_i = \frac{1 - F - \left(\dfrac{\gamma(p-t)}{v^\gamma}\right)^{\frac{1}{1-\gamma}}}{p_{B+}t_B} - \frac{\varepsilon}{\beta}.$$

Substituting these into equation (3.1) defines the indirect utility function:

$$U_i^* = \left(\frac{1 - F - \left(\frac{\gamma(p-t)}{v^\gamma} \right)^{\frac{1}{1-\gamma}}}{p_B + t_B} \right)^\beta \left(\frac{\varepsilon(p_B + t_B)}{\beta} \right)^\alpha. \tag{3.4}$$

Traditional Farmers

The farmers using traditional technology have preferences identical to those using modern technology. The only difference is that they pay no fixed costs and thus use a less efficient technology $y_j = W_j^\eta$, with $0 < \eta < \gamma < 1$. The traditional farmer's maximization problem is given by equation (3.5):

$$Max \ U_j = B_j^\beta (\delta_j y_j)^\alpha \tag{3.5}$$

subject to

$$(1 - \delta_j)(p - t)W_j^\eta = (p_B + t_B)B_j + vW_j,$$

which by an analogous procedure yields the indirect utility function:

$$U_j^* = \left(\frac{1 - \left(\frac{\eta(p-t)}{v^\eta} \right)^{\frac{1}{1-\eta}}}{p_B + t_B} \right)^\beta \left(\frac{\varepsilon(p_B + t_B)}{\beta} \right)^\alpha. \tag{3.6}$$

Technology Choice

Given these production and utility levels, in stage one farmers will (or will not) switch from subsistence farming to modern farming if $U_i^* > (<)U_j^*$. Substituting from equations (3.4) and (3.6) and rearranging gives the result in equation (3.7):

$$U_i^* - U_j^* < 0 \text{ if } \psi < F \tag{3.7}$$

in which $\psi \equiv \left(\frac{\gamma(p-t)}{v^\gamma} \right)^{\frac{1}{1-\gamma}} - \left(\frac{\eta(p-t)}{v^\eta} \right)^{\frac{1}{1-\eta}}.$

Suppose next that farmers are situated at a range of locations with differing transport costs $t_h \in [t_0,....,\tilde{t},...T]$, in which $t_0 < \tilde{t} < T$. Define $\tilde{t}$ such that for some $\tilde{F}>0$, $\tilde{\psi}=\left(\dfrac{\gamma\left(p-\tilde{t}\right)}{v^{\gamma}}\right)^{\frac{1}{1-\gamma}}-\left(\dfrac{\eta\left(p-\tilde{t}\right)}{v^{\eta}}\right)^{\frac{1}{1-\eta}}=\tilde{F}$. Observe that at

$\tilde{\psi}$ farmers are indifferent between the technologies. The following results are obtained:

Result 1: At locations $t_h<\tilde{t}$, farmers will find it profitable to switch to the improved technology, whereas at locations with $t_h<\tilde{t}$, farmers will remain with the old technology.

Proof: Note that $\dfrac{d\psi}{dt}=-\dfrac{1}{1-\gamma}\left(\dfrac{\gamma\left(p-t\right)}{v^{\gamma}}\right)^{\frac{1}{1-\gamma}}+\dfrac{1}{1-\eta}\left(\dfrac{\eta\left(p-t\right)}{v^{\eta}}\right)^{\frac{1}{1-\eta}}<0$ since

$\eta<\gamma$ by construction. Since Ψ is continuous in t, then from the intermediate value theorem it follows that for locations where $t_h<\tilde{t}$, then $\Psi>\tilde{F}$ and these farmers switch to the improved technology, whereas in locations with $t_h>\tilde{t}$, then $\Psi<\tilde{F}$ and it pays to remain with the old technology.

Let $E_k=\dfrac{\partial y_k}{\partial t}\dfrac{t}{y_k}$, $(k=i,j)$ be the output elasticity of demand with respect to transport costs of goods sold in markets.

Result 2: The output of farmers who adopt more modern, improved technologies is more responsive to changes in transport costs than that of farmers who use the old technology (that is, $|E_i|>|E_j|$).

Proof: Note that since $y_i=\left(\dfrac{\gamma\left(p-t\right)}{v}\right)^{\frac{\gamma}{1-\gamma}}$ then $E_i=\dfrac{-\gamma t}{\left(p-t\right)\left(1-\gamma\right)}$ and by analogy $E_j=\dfrac{-\eta t}{\left(p-t\right)\left(1-\eta\right)}$. Since by assumption $\gamma>\eta$, it follows that $|E_i|>|E_j|$.

An implication of this result is that the farmers with the improved technology will be better integrated into output markets and will sell a greater fraction of their output.

Corollary: Farmers with improved technology (modern farmers) consume a smaller fraction of their output and sell a greater proportion of their output on the market.

Proof: Using equation (3.3) it is evident that $\dfrac{d\delta_i}{d\gamma}<0$.

The analysis therefore suggests three hypotheses. First, the model suggests that adoption of new technologies will be more pervasive where transport costs are lower (Result 1). Next, the model predicts that reductions in transport costs will have a larger impact on the marketed output of farmers using modern farming techniques (Result 2). Finally, farmers with more modern technologies are likely to be better integrated into markets.

Annex 3B Spatial Production Allocation Model Regression Results

Table 3B.1 Yams

	Dependent variable: ln(yams high-input production)		Dependent variable: ln(yams low-input production)	
	(1)	(2)	(3)	(4)
	OLS	2SLS	OLS	2SLS
ln(transport cost to market)	−0.232**	−0.310***	−0.0300	0.0849
	(−2.47)	(−2.81)	(−0.26)	(0.64)
ln(population)	0.627**	0.651**	0.922***	0.839**
	(2.18)	(2.26)	(2.65)	(2.42)
ln(population)^2	−0.0358*	−0.0388**	−0.0429*	−0.0343
	(−1.95)	(−2.10)	(−1.93)	(−1.54)
ln(distance to mine)	−0.159	−0.149	−0.378***	−0.351***
	(−1.53)	(−1.42)	(−3.02)	(−2.80)
ln(high-input potential)	−2.844***	−2.745***		
	(−5.38)	(−5.16)		
ln(high-input potential)^2	0.245***	0.237***		
	(6.38)	(6.12)		
ln(low-input potential)			0.481*	0.541**
			(1.83)	(2.07)
ln(low-input potential)^2			−0.000466	−0.00764
			(−0.02)	(−0.29)
Observations	8,783	8,684	8,747	8,649

Sources: SPAM (HarvestChoice 2012); LandScan 2006; FAO and IIASA 2000; and calculations.
Note: 2SLS = two-stage least squares; OLS = ordinary least squares. *t*-statistics in parentheses.
***$p < 0.01$, **$p < 0.05$, *$p < 0.10$.

Table 3B.2 Rice

	Dependent variable: ln(yams high-input production)		Dependent variable: ln(yams low-input production)	
	(1)	(2)	(4)	(5)
	OLS	2SLS	OLS	2SLS
ln(transport cost to market)	−0.357***	−0.528***	0.348***	0.522***
	(−3.18)	(−3.96)	(2.98)	(3.79)
ln(population)	0.567*	0.596*	0.0742	−0.0523
	(1.79)	(1.87)	(0.22)	(−0.16)
ln(population)^2	−0.0383*	−0.0429**	0.0113	0.0249
	(−1.86)	(−2.06)	(0.52)	(1.15)

(continued next page)

Table 3B.2 (continued)

	Dependent variable: ln(yams high-input production)		Dependent variable: ln(yams low-input production)	
	(1)	(2)	(4)	(5)
	OLS	2SLS	OLS	2SLS
ln(distance to mine)	−0.199	−0.175	−0.577***	−0.514***
	(−1.64)	(−1.43)	(−4.56)	(−4.04)
ln(high-input potential)	−3.236***	−3.252***		
	(−10.07)	(−10.05)		
ln(high-input potential)^2	0.349***	0.352***		
	(13.40)	(13.37)		
ln(low-input potential)			2.702***	2.792***
			(8.76)	(9.03)
ln(low-input potential)^2			−0.115***	−0.122***
			(−3.74)	(−3.94)
Observations	6,548	6,449	6,778	6,679

Sources: SPAM (HarvestChoice 2012); LandScan 2006; FAO and IIASA 2000; and calculations.
Note: 2SLS = two-stage least squares; OLS = ordinary least squares. *t*-statistics in parentheses.
***$p < 0.01$, **$p < 0.05$, *$p < 0.10$.

Table 3B.3 Cassava

Dependent variable: ln(cassava low-input production)	(1)	(2)
	OLS	2SLS
ln(transport cost to market)	−0.323**	−0.243
	(−2.02)	(−1.29)
ln(population)	1.124**	1.011**
	(2.42)	(2.17)
ln(population)^2	−0.0625**	−0.0526*
	(−2.09)	(−1.75)
ln(distance to mine)	−0.750***	−0.685***
	(−4.34)	(−3.94)
ln(low-input potential)	1.632***	1.729***
	(5.15)	(5.46)
ln(low-input potential)^2	−0.0293	−0.0383
	(−0.96)	(−1.25)
Observations	7,533	7,435

Sources: SPAM (HarvestChoice 2012); LandScan 2006; FAO and IIASA 2000; and calculations.
Note: 2SLS = two-stage least squares; OLS = ordinary least squares. *t*-statistics in parentheses.
***$p < 0.01$, **$p < 0.05$, *$p < 0.10$.

Table 3B.4 Maize

Dependent variable: ln(maize low-input production)	(1) OLS	(2) 2SLS
ln(transport cost to market)	0.885***	1.212***
	(6.27)	(7.31)
ln(population)	1.350***	1.263***
	(3.09)	(2.87)
ln(population)^2	−0.0691**	−0.0583**
	(−2.48)	(−2.07)
ln(distance to mine)	−1.863***	−1.866***
	(−11.95)	(−11.86)
ln(low-input potential)	1.906***	2.336***
	(2.87)	(3.42)
ln(low-input potential)^2	0.0148	−0.0171
	(0.26)	(−0.30)
Observations	9,242	9,144

Sources: SPAM (HarvestChoice 2012); LandScan 2006; FAO and IIASA 2000; and calculations.
Note: 2SLS = two-stage least squares; OLS = ordinary least squares. *t*-statistics in parentheses.
***$p < 0.01$, **$p < 0.05$, *$p < 0.10$.

Table 3B.5 Summary Statistics: SPAM

Variable	Observations	Mean	Standard deviation	Minimum	Maximum
Subsistence production					
Yams (tons)	10,015	1,072.397	2,256.246	0	29,988.5
Rice (tons)	10,015	93.25695	136.4678	0	1,019.2
Cassava (tons)	10,015	3,419.833	8,387.33	0	143,613.1
Maize (tons)	10,015	491.527	854.9417	0	10,103.8
High-input production					
Yams (tons)	10,015	1,115.142	7,199.12	0	179,121.0
Rice (tons)	10,015	196.0147	1,249.861	0	31,679.5
Low-input potential yield					
Yams (kg/ha)	10,015	609.166	383.2807	0	1,747
Rice (kg/ha)	10,015	494.7672	436.5076	0	1,792
Cassava (kg/ha)	10,015	833.5371	681.5004	0	2,775
Maize (kg/ha)	10,015	1,209.439	625.5546	0	3,556
High-input potential yield					
Yams (kg/ha)	10,015	3,253.258	1,780.383	0	7,028
Rice (kg/ha)	10,015	1,646.77	1,532.992	0	6,389

(continued next page)

Table 3B.5 (continued)

Variable	Observations	Mean	Standard deviation	Minimum	Maximum
Other controls					
Population	10,015	13,639.3	44,454.81	0	1.639 million
Market transport cost (US$)	10,015	7.989746	4.76409	0.1	38.2
Natural path (hours)	10,015	20.98574	13.42766	0	81.02
Distance to mine (km)	10,003	189.7956	110.7933	2.9	499.4

Sources: SPAM (HarvestChoice 2012); LandScan 2006; FAO and IIASA 2000; and calculations.
Note: kg/ha = kilogram per hectare; SPAM = Spatial Production Allocation Model.

Annex 3C Living Standards Measurement Study Regression Results

Table 3C.1 Instrumental Variable–Probit Estimates

Dependent variable: Dummy = 1 if uses machinery	Exclusion Restriction 1 (nonagricultural income) (1)	Exclusion Restriction 2 (neighborhood effects) (2)
ln(transport cost to market)	−0.177***	−0.254***
	(−2.33)	(−3.28)
ln(nonagricultural income)	0.037**	
	(1.98)	
ln(average machine use in village)		4.814***
		(31.86)
Land	0.015***	0.005***
	(5.19)	(2.43)
Age of household head	−0.013	−0.022
	(−0.89)	(−1.54)
Age^2	0.000	0.000
	(1.14)	(1.69)
Dummy = 1 if head is literate	0.010	0.075
	(0.12)	(0.97)
Fertilizer purchased (kilograms)	−0.000	0.001
	(−0.17)	(0.44)
Distance to mine	0.123**	0.220***
	(2.12)	(3.51)
First-stage estimates	0.639***	0.632***
ln(natural path)	(48.63)	(51.95)
Observations	1,354	1,354

Sources: Nigeria Living Standards Measurement Study–Integrated Surveys on Agriculture (National Bureau of Statistics 2010); and calculations.
Note: Robust t-statistics in parentheses.
***p < 0.01, **p < 0.05, *p < 0.10.

Table 3C.2 Instrumental Variable Estimates with Heckman Selection Correction

Dependent variable: ln(crop revenue)	(1) IMR$_1$ (nonagricultural income)	(2) IMR$_2$ (neighborhood effects)
ln(transport cost to market)	−2.200***	−2.455***
	(−3.43)	(−6.54)
Land	0.053*	0.048***
	(1.73)	(4.47)
Age of head	0.181**	0.094
	(2.14)	(1.51)
Age^2	−0.001**	−0.001
	(−1.96)	(−1.28)
Dummy = 1 if head is literate	1.595***	0.614
	(2.92)	(1.44)
Fertilizer purchased	−0.016	−0.012
	(−1.17)	(−0.97)
ln(distance to mine)	−0.360	−0.620
	(−0.62)	(−1.61)
Dummy = 1 if plot is irrigated	−0.696	0.049
	(−0.72)	(0.06)
Household agricultural labor	0.386	0.471**
	(1.37)	(2.14)
Labor^2	−0.049**	−0.048***
	(−2.00)	(−2.59)
Dummy = 1 if AEZ is warm/subhumid	−0.817	−0.828
	(−1.25)	(−1.49)
Dummy = 1 if AEZ is warm/humid	1.687	−0.629
	(0.47)	(−0.17)
Dummy = 1 if AEZ is cool/subhumid	2.283**	1.670
	(1.98)	(1.58)
IMR$_1$	−0.413	
	(−0.12)	
IMR$_2$		−0.193
		(−0.56)
Constant	3.806	7.701***
	(0.67)	(2.66)
First-stage results		
ln(natural path)	0.699***	0.687***
	(42.01)	(67.08)
Angrist-Pischke test of weak identification	1,764.72	4,499.78
	P = 0.0000	P = 0.0000
Observations	393	621

Sources: Nigeria Living Standards Measurement Study–Integrated Surveys on Agriculture (National Bureau of Statistics 2010); and calculations.
Note: AEZ = Agro-Ecological Zone; IMR = inverse Mills ratio. Robust *t*-statistics in parentheses.
***$p < 0.01$, **$p < 0.05$, *$p < 0.10$.

Table 3C.3 Instrumental Variable Estimates, Heterogeneous Effects

Dependent variable: ln(crop revenue)	(1) Full sample	(2) Modern	(3) Traditional
ln(transport cost to market)	−1.774***	−2.441***	−1.422***
	(−8.44)	(−6.57)	(−5.47)
Land	0.031***	0.049***	0.018**
	(4.38)	(4.66)	(2.37)
Age of head	0.033	0.094	0.015
	(0.98)	(1.51)	(0.39)
Age^2	−0.000	−0.001	−0.000
	(−0.76)	(−1.29)	(−0.29)
Dummy = 1 if head is literate	0.618***	0.634	0.533***
	(3.34)	(1.49)	(2.62)
Fertilizer purchased	−0.005	−0.012	−0.000
	(−0.62)	(−0.98)	(−0.02)
ln(distance to mine)	−0.351**	−0.609	−0.280
	(−2.09)	(−1.60)	(−1.45)
Dummy = 1 if plot is irrigated	0.781	0.046	1.189*
	(1.55)	(0.05)	(1.90)
Household agricultural labor	0.322***	0.472**	0.337***
	(3.22)	(2.17)	(2.77)
Labor^2	−0.041***	−0.048***	−0.047***
	(−4.07)	(−2.59)	(−3.29)
Dummy = 1 if AEZ is warm/subhumid	0.664***	−0.737	0.960***
	(2.81)	(−1.45)	(3.39)
Dummy = 1 if AEZ is warm/humid	−0.201	−0.677	0.279
	(−0.51)	(−0.18)	(0.67)
Dummy = 1 if AEZ is cool/subhumid	3.362***	1.830*	1.504
	(3.89)	(1.83)	(0.49)
Constant	5.491***	7.393***	4.843***
	(4.15)	(2.65)	(3.23)
First-stage results			
ln(natural path)	0.632***	0.685***	0.607
	(53.06)	(67.94)	(37.68)
Angrist-Pischke test of weak identification	2,814.98	4,616.49	1,420.08
	P = 0.0000	P = 0.0000	P = 0.0000
Observations	2,598	624	1,974

Sources: Nigeria Living Standards Measurement Study–Integrated Surveys on Agriculture (National Bureau of Statistics 2010); and calculations.
Note: AEZ = Agro-Ecological Zone. Robust t-statistics in parentheses.
***p < 0.01, **p < 0.05, *p < 0.10.

Table 3C.4 Summary Statistics from the Living Standards Measurement Study–Integrated Surveys on Agriculture

Variable	Full sample		Mechanized only		Traditional only	
	Observations	Mean	Observations	Mean	Observations	Mean
Crop revenue (US$)	3,022	115.43	764	218.12	2,255	80.66
Transport cost to market (US$)	2,607	5.10	624	5.49	1,983	4.97
Nonagricultural income	2,986	461.38	474	225.74	1,134	309.77
Neighborhood machinery use	3,068	0.25	770	0.70	2,298	0.10
Dummy = 1 if mechanized	3,076	0.25	773	1.00	2,303	0
Land of household (km²)	3,087	10.68	773	16.33	2,303	8.81
Age of household head	4,978	49.56	773	49.85	2,295	50.89
Dummy = 1 if literate	4,982	0.63	773	0.53	2,300	0.53
Dummy = 1 if irrigates land	3,085	0.04	773	0.06	2,302	0.03
Fertilizer purchased (kilograms)	3,034	1.28	764	1.45	2,269	1.22
Household agricultural labor	4,991	1.31	773	2.22	2,301	1.99
Distance to mine (km)	2,607	139.95	624	155.33	2,255	80.66

Sources: Nigeria Living Standards Measurement Study–Integrated Surveys on Agriculture (National Bureau of Statistics 2010); and calculations.

Notes

1. However, there is some evidence that fertilizer use has increased recently in some African countries (Sheahan and Barrett 2014).
2. For a recent study of agricultural intensification in Africa, see Binswanger-Mkhize and Savastano (2014).
3. For an early and concise summary see Feder, Just, and Zilberman (1985).
4. The SPAM data set disaggregates crop production statistics into 10 kilometer x 10 kilometer cells throughout all of Nigeria. This data set is useful because it clearly distinguishes between production of various crops using different input systems (labeled low input and high input, with the former referred to as "subsistence/low input" and the latter referred to as "irrigated," although this latter category encompasses mechanized and all input-intensive forms of agriculture). For instance, subsistence input systems are defined as "rainfed crop production which uses traditional varieties and mainly manual labor without (or with little) application of nutrients or chemicals for pest and disease control." Irrigated input systems refer to production in which "the crop area [is] equipped with either full or partial control

irrigation . . . [normally using] . . . high level of inputs such as modern varieties and fertilizer as well as advanced management such as soil/water conservation measures" (http://mapspam.info/methodology/#production_systems).

5. See FAOSTAT: http://faostat.fao.org/site/339/default.aspx.

6. For a more detailed analysis of these results, as well as the full regression tables, see annex 3B.

7. Although this would be interesting to test empirically, it is difficult to test this using SPAM data. To do so, one would have to aggregate all crops into a single-yield metric. The natural way to do this would be to use crop prices to calculate production values. However, data on local crop prices in Nigeria are very difficult to obtain, especially for the number of crops and number of local markets that would be required.

8. Because richer areas (those with higher nonagricultural income) are more likely to use machinery, nonagricultural income and average machinery use are likely to be highly correlated. Thus, the two are included separately.

9. Even renting space in a truck to carry goods to market is a lumpy cost. Hayami and Kawagoe (1993) note that the transport cost in Indonesia is R5 rupees per kilogram or higher for a load up to 200 kilograms by pony wagon, but that the cost declines to R2.5 rupees per kilogram if a two-ton load is carried by a small truck.

10. For a discussion of coordination failures and the big-push model, see, for example, Rosenstein-Rodan (1943); Murphy, Shleifer, and Vishny (1989); and Hoff (2000).

References

Binswanger-Mkhize, Hans, and Sara Savastano. 2014. "Agricultural Intensification: The Status in Six African Countries." Policy Research Working Paper 7116, World Bank, Washington.

Conley, Timothy G., and Christopher R. Udry. 2010. "Learning about a New Technology: Pineapple in Ghana." *American Economic Review* 100 (1): 35–69.

FAO and IIASA (Food and Agriculture Organization of the United Nations and the International Institute for Applied Systems Analysis). 2000. "Global Agro-Ecological Zones (GAEZ)." FAO, Rome; and IIASA, Laxenberg, Austria.

Feder, Gershon, Richard E. Just, and David Zilberman. 1985. "Adoption of Agricultural Innovations in Developing Countries: A Survey." *Economic Development and Cultural Change* 33 (2): 255–99.

Fuglie, Keith O., Sun Ling Wang, and V. Eldon Ball, eds. 2012. *Productivity Growth in Agriculture: An International Perspective.* Oxfordshire, UK: CAB International.

Gollin, Douglas, Michael Morris, and Derek Byerlee. 2005. "Technology Adoption in Intensive Post-Green Revolution Systems." *American Journal of Agricultural Economics* 87 (5): 1310–16.

HarvestChoice. 2012. "Spatial Allocation of Agricultural Production." International Food Policy Research Institute, Washington, DC, and University of Minnesota, St. Paul, MN. http://harvestchoice.org/node/2248.

Hayami, Yujiro, and Toshihiko Kawagoe. 1993. *The Agrarian Origins of Commerce and Industry: A Study of Peasant Marketing in Indonesia.* New York: St. Martin's Press.

Heckman, James J. 1976. "The Common Structure of Statistical Models of Truncation, Sample Selection and Limited Dependent Variables and a Simple Estimator for Such Models." *Annals of Economic and Social Measurement* 5 (4): 475–92.

Hoff, Karla. 2000. "Beyond Rosenstein-Rodan: The Modern Theory of Underdevelopment Traps." Working Paper 28725, World Bank, Washington, DC.

LandScan. 2006. Global Population Database (2006 release). Oak Ridge National Laboratory, Oak Ridge, Tennessee. http://www.ornl.gov/landscan/.

Larson, Donald F., and Mauricio León. 2006. "How Endowments, Accumulations, and Choice Determine the Geography of Agricultural Productivity in Ecuador." *World Bank Economic Review* 20 (3): 449–71.

Mundlak, Yair. 1988. "Endogenous Technology and the Measurement of Productivity." In *Agricultural Productivity: Measurement and Explanation,* edited by Susan M. Dapalbo and John M. Antle. Washington, DC: International Food Research Policy Institute.

Mundlak, Yair, Rita Butzer, and Donald Larson. 2012. "Heterogeneous Technology and Panel Data: The Case of the Agricultural Production Function." *Journal of Development Economics* 99 (1): 139–49.

Mundlak, Yair, Donald Larson, and Rita Butzer. 1999. "Rethinking within and between Regressions: The Case of Agricultural Production Functions." *Annales d'Economie et Statistique* 55–56 (Sep–Dec): 475–501.

Murphy, Kevin M., Andrei Shleifer, and Robert W. Vishny. 1989. "Industrialization and the Big Push." *Journal of Political Economy* 97 (5): 1003–26.

National Bureau of Statistics, Federal Republic of Nigeria. Nigeria General Household Survey (GHS), Panel 2010, Ref. NGA_2010_GHS_v02_M. Data set downloaded from http://econ.worldbank.org/WBSITE/EXTERNAL/EXTDEC/EXTRESEARCH/EXTL SMS/0,,contentMDK:22949589~menuPK:4196952~pagePK:64168445~piPK:641683 09~theSitePK:3358997,00.html.

Pinstrup-Andersen, Per, Rajul Pandya-Lorch, and Mark W. Rosegrant. 1997. "The World Food Situation: Recent Developments, Emerging Issues, and Long-Term Prospects." International Food Policy Research Institute, Washington, DC.

Rosenstein-Rodan, Paul. 1943. "Problems of Industrialization of Eastern and Southeastern Europe." *Economic Journal* 53 (June–September): 202–11.

Schiff, Maurice, and Claudio E. Montenegro. 1997. "Aggregate Agricultural Supply Response in Developing Countries: A Survey of Selected Issues." *Economic Development and Cultural Change* 45 (2): 393–410.

Sheahan, Megan, and Christopher B. Barrett. 2014. "Understanding the Agricultural Input Landscape in Sub-Saharan Africa: Recent Plot, Household, and Community-Level Evidence." Policy Research Working Paper 7014, World Bank, Washington, DC.

Slootmaker, Chris. 2013. "Technology Adoption, Risk, and Intrahousehold Bargaining in Subsistence Agriculture." Working Paper, Colorado State University, Fort Collins.

World Bank. 2012. "Agriculture, Policy Reform and Poverty in Sub Saharan Africa." Discussion Paper 280, World Bank, Washington, DC.

———. 2013. "Unlocking Africa's Agricultural Potential: An Action Agenda for Transformation." Africa Region Sustainable Development Series, World Bank, Washington, DC.

Chapter **4**

Role of Transport Infrastructure in Conflict-Prone and Fragile Environments: Evidence from the Democratic Republic of Congo

Introduction

In conflict-prone situations, access to markets is deemed necessary to restore economic growth and generate the preconditions for peace and reconstruction. Hence, the rehabilitation of damaged transport infrastructure has emerged as an overarching investment priority among donors and governments in conflict-prone and fragile states. The New Economic Partnership for Africa's Development (NEPAD) has proposals for nine highways across the continent, at an estimated cost of US$4.2 billion (African Development Bank 2003), all of which pass through fragile states (as defined by the Organisation for Economic Co-operation and Development). In most cases, the portions of the roads that need the most rehabilitation lie within these conflict-prone countries. In Afghanistan, the U.S. Agency for International Development provided more than US$1.8 billion between 2002 and 2007 to reconstruct roads (GAO 2008), while the U.S. Department of Defense has allocated about US$300 million in Commander's Emergency Response Program funds for roads. There is, however, little empirical evidence on the direct causal impact of access to markets on well-being in fragile situations when the risks of renewed conflict are high, and even less evidence on the combined impact of transport costs and conflict.

Why the DRC?

The Democratic Republic of Congo (DRC), with its history of periodic conflict and economic stagnation, provides an apt case study for obtaining empirical evidence on the effects of transport infrastructure in the context of conflict.

Table 4.1 Road Conditions in the DRC versus Low-Income-Country Average

Indicator	Units	Low-income-country average	Democratic Republic of Congo
Paved road density	km/1,000 km² of land	16	1
Unpaved road density	km/1,000 km² of land	68	14
Paved road traffic	Average daily traffic	1,028	257
Unpaved road traffic	Average daily traffic	55	20
Perceived transport quality	Percentage of firms identifying as major business constraint	23	30

Source: World Bank 2010.

Constant conflict, poor governance, and lack of infrastructure have left the DRC one of the poorest countries in the world, with the average Congolese resident living on less than US$0.75 per day. And yet, geography and natural endowments of unexploited mineral wealth (estimated value US$24 trillion [UNEP 2011]) and 22.5 million hectares of uncultivated, unprotected, nonforested fertile land give the DRC the potential to become one of the richest countries in the region. Harnessing the growth potential of these endowments is not without challenges, partially due to severely deficient infrastructure (only four provincial capitals out of ten can be reached by road from the national capital, Kinshasa), even by the standards of other low-income countries (table 4.1). Infrastructure improvements will undoubtedly need to include significant road improvement and construction projects in the DRC, where spending on transport infrastructure was approximately US$230 million per year during the mid-2000s, and increased to US$275 million per year in 2008 and 2009 (Pushak and Briceño-Garmendia 2011). However, the conflict still erupting in parts of the country raises the question of whether improvements in transport infrastructure will bring benefits to local economies. Such interventions could also have perverse effects and provide violent militias easier access to the entire country.

Challenges in Estimating the Effect of Transport Cost and Conflict

The objective of this chapter is to shed light on how transport costs and conflict are interlinked and how they directly and jointly affect welfare indicators. Lower transport costs and better access to markets potentially encourages production and trade. Conflict near rural households has the potential to disrupt agricultural production. Additionally, conflict near markets and along the roads can disrupt economic activity and diminish people's willingness to travel and trade.

Violent conflicts are therefore likely to have a considerable impact on public and private investment, and consequently on economic conditions and the overall welfare of households. One of the challenges involved in this research and in estimating the effects of transport cost and conflicts is that it can be difficult to precisely measure the effects at a fine spatial level. Conflicts occur and are recorded at point sources, but the effect of conflict is likely to be widespread in the surrounding area. And as discussed in previous chapters, once transport cost is accurately measured (which is no small task), estimating the effect of reducing transport cost on welfare remains a challenge because of the nonrandom placement of transport infrastructure. Roads may be placed near areas of high economic potential or in poorer areas to enable residents of those areas to improve their economic condition through better access to markets. This nonrandom placement makes it difficult to infer whether roads near wealthier areas caused those areas to be richer, or whether it was the other way around (or, alternatively, whether roads near poor areas are ineffective, or if they were built there precisely because the area is poor). Similarly, determining what welfare would have been in the absence of roads is difficult. Failure to take this endogeneity into account in estimating the benefits of roads can lead to statistically biased results. At the same time, conflict is also endogenous given that it is closely related to wealth: conflict negatively affects wealth, but by the same token, low levels of income could trigger incidences of conflict, and this reverse causality needs to be accounted for when estimating the effect of conflict so that consistent estimates can be obtained.

To motivate and contextualize the subsequent empirical analysis, this chapter first presents a simple theoretical model, outlined in box 4.1, to analyze how transport costs and conflict are interlinked and how they directly and jointly affect welfare indicators.

BOX 4.1

Theoretical Framework

Structure of the Model

The model assumes that the economy consists of a continuum of individuals who are uniformly distributed along the unit interval $n_i \in [0,1]$, where the individual at location $n_i = 0$ is closest to the market and that at $n_i = 1$ is farthest away from the market. Transport costs to the market are given by $t_i = z n_i$, $z > 0$. Individuals at each location can either choose to farm or join a rebel force that loots from those who farm.

(continued next page)

Box 4.1 (continued)

The model is set up as a stage game in which the rebels have a strategic advantage over farmers (that is, move before farmers).

- First, individuals choose whether to farm or to join the rebel group and loot.

- Given the set of rebels (or equivalently the set of farmers at known locations), in the second stage the rebels determine effort levels and a looting strategy that consists of the decision to attack either at the market or where the farmers are located.

- In the third stage, farmers determine the types of goods to produce and the production levels of each. There are two types of produced goods—those produced for sale in the markets, which must incur transport costs, and goods for domestic consumption (subsistence farming). Both products are potentially vulnerable to theft by rebels.

Predictions of the model

The first result of the model indicates that lower transport costs will induce a switch from farming to joining the rebels (and vice versa) when the costs of stealing marketed goods are sufficiently low (high). The intuition behind this result is that lower transport costs make production for the market more attractive for the farmers and therefore, when the costs of looting from the markets are sufficiently low, the strategic advantage (that is, payoffs) accruing to the rebels increases because they are able to loot a greater share of the aggregate output produced by farmers. As a result, payoffs available to the rebels from theft of the higher aggregate output can rise faster than the increased output of the marginal farmer, which will induce a switch from farming to joining the rebels. A corollary of result 1 is that when more households join the rebels, agricultural output would decline, all else equal, so there can be no presumption that transport cost reductions induce the desired outcomes.

Result 2 suggests that the consequences of increased looting at each location depend critically upon how much weight in the utility of individuals is put on the marketed good relative to the subsistence good. When the marketed good is given sufficient weight relative to the subsistence product, the marginal welfare costs of theft at the market are higher and vice versa. An implication of this result is that the combined effects of simultaneous changes in transport costs and conflict are also ambiguous even in this highly stylized framework.

Innovations and Contributions of This Chapter

This chapter uses some novel data sets to analyze the effect of transport costs to the nearest market (defined as cities with populations of at least 50,000) and conflict on several indicators of welfare—a wealth index, a multidimensional poverty indicator (constructed from the Demographic and Health Survey 2007 for the DRC) and local GDP for 2006.[1] The household indicators are useful because they provide insight into the explicit household-level effects of reducing transport costs while controlling for specific heterogeneities of the households. The local GDP data provide insights into the level of economic activity at specific locations.

To conduct the analysis, a thorough geographic information vector data set on the transportation network of major trunk roads as well as rural roads throughout the country was obtained from DeLorme.[2] Combining this data set with African Infrastructure Country Diagnostic data (which includes data on road quality attributes) and using the Highway Development Management Model,[3] this study computes the cost per kilometer of transporting a ton of goods in a heavy truck from every location within the DRC to the closest market.

To correct for any placement bias inherent in estimating the benefits of transportation infrastructure expenditures, this study uses an instrumental variables (IVs) estimation approach, and generates an innovative instrument called the natural-historical path (NHP). As suggested by its name, the NHP takes into consideration historical data on caravan routes from the nineteenth century, as well as the terrain and historical land cover within the DRC, to estimate the quickest path, on foot, to and from anywhere within the DRC's borders. Note that the NHP is an enhanced version of the natural path IV used in previous chapters. Specifically, the NHP augments the natural path with the inclusion of historical caravan routes.[4]

The literature on the economic benefits of roads typically relies on one of two types of IVs: straight-line IVs or historical path IVs. Even though these two types of IVs are very different in their formulation, they both attempt to estimate the natural way for humans to travel over land, in the absence of a road network. However, as discussed in chapter 2, straight-line IVs have been criticized because they fail to account for topography of the land. Historical paths are useful as IVs in that they represent the easiest path to travel over land. Having been constructed with little or no technology, they generally follow smoother terrain and have been used for hundreds of years, and are thus the most cost-effective walking route. At the same time, they are usually not correlated with the current economic benefits that lead to the endogeneity bias, given that in many cases these routes were identified well over a century ago.[5]

A combination of both the natural path and historical caravan data yields an improved estimate of how people traveled over land in previous centuries, and thus is the best possible IV for transport cost. A major problem with using only historical path data is that people now live in areas that may have been uninhabited, or not a part of the trade network, many years ago. Historical path data will therefore not be able to identify the likely paths that would have been used to travel to and from those newer locations. By using natural path data, this study is able to fill in gaps in the historical caravan data to get a more complete picture of travel paths.[6]

To examine conflict, the study uses the Armed Conflict Location Events Dataset (ACLED) (Raleigh et al. 2010) version 4, which reports information on the location, date, and other characteristics of politically violent events for all countries on the African continent during 1997–2013. Using the ACLED in its raw point format brings up a host of technical and methodological problems. First, each conflict is pinned to a single geographic point, and does not capture the effects of conflict on the surrounding area. For instance, battles may have been fought over a large area, and the effects will be felt by an area significantly larger than a solitary point. Second, conflict points cannot capture conflict intensity (for instance, one isolated conflict point versus a cluster of conflict points, or one small conflict with one fatality versus a major battle with thousands). Finally, ACLED is subject to some geographic imprecision resulting from how the data were obtained (for instance, conflicts occurring in rural areas are sometimes allocated to the nearest village).[7]

An innovative methodology, a kernel density function, is used to measure conflict around households and markets. This technique transforms conflict points into a smooth surface and generalizes a set of conflict locations. To calculate the value at any point, the kernel density function takes a weighted average of all the conflicts around that point to create the surface. The magnitude of the weight declines with distance from the point, according to the chosen kernel function. Box 4.2 provides information on the kernel estimation technique. Panel a of map B4.2.1 in box 4.2 shows the original ACLED conflict data and panel 2 shows the conflict density map that was estimated using this technique. For a more detailed explanation of the data innovations described above, see appendix A on Geospatial Analysis.

Using these data, several measures of conflict are constructed. The first is the "kernelly" estimated number of fatalities in the five years preceding the Demographic and Health Survey (DHS) data set (2003–07) and local GDP data set (2002–06). This variable is calculated around each household and also around each market. A dummy variable is generated that indicates whether relatively high levels of conflict have occurred near households and markets. The dummy variable takes a value of 1 if the kernelly estimated fatalities are

Conflict Measure Using Kernel Estimation

The kernel density interpolation technique is used to generalize unique incident locations into an area. It involves placing a symmetrical surface over each point (incident), evaluating the distance from the point to a reference location (bandwidth or search radius) based on a mathematical function (kernel function), and summing the value of all the surfaces for that reference location (pixels of 10 kilometers × 10 kilometers).

For instance, if a conflict occurs exactly on the point that is being calculated, the value of that conflict will receive a weight of 1. A conflict that is 5 kilometers away from the point will receive a weight of α and a conflict 10 kilometers away will receive a weight of β, where $1 > \alpha > \beta > 0$. Eventually, at some distance, referred to as the bandwidth, the weight becomes zero. This procedure is repeated for all reference locations.

For more information on how the kernel function and bandwidth were chosen, see appendix A.

Map B4.2.1 Estimated Conflict Surface

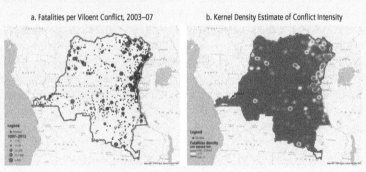

a. Fatalities per Viloent Conflict, 2003–07   b. Kernel Density Estimate of Conflict Intensity

Sources: Raleigh et al. 2010; and calculations.

greater than the median number of fatalities due to violent conflict near both households and the closest market.[8]

Conflict is another variable that, if not treated appropriately in a statistical regression, can lead to biases due to reverse causality. To account for the endogeneity of conflicts, social fractionalization indices around households and markets are used as instruments. Social fractionalization, which measures religious, ethnic, and linguistic diversity within a country or region, can be used as a valid instrument for conflict. Earlier studies have found ethnolinguistic fractionalization to be a strong determinant of the probability and duration of conflict (Wegenast and Basedau 2014; Esteban, Mayoral, and Ray 2012; Esteban and Schneider 2008; Schneider and Wiesehomeier 2008; Montalvo and Reynal-Querol 2005; Reynal-Querol 2002; Gurr 2000; Collier and Hoeffler 1998; Horowitz 1985). The validity of the instrument is based on the observed correlation between higher levels of fractionalization and levels of conflict.[9] The fractionalization variable used in this study satisfies the exclusion restriction, that is, causality could not run from welfare to fractionalization, given that the fractionalization variable is generated using demographic data published in 2001 (details of the methodology used to generate this variable are provided later in this section) while the income measures used are from 2006 and 2007. Income measures in these years cannot affect predetermined, and hence exogenous, ethnic fractionalization. The level of conflict is also instrumented by the distance to the eastern border because this area has higher levels of conflict relative to the rest of the country, and distance should not directly affect income, suggesting that this is a reasonable IV.

Micro-level ethnic fractionalization measures are used, that is, fractionalization within a 50 kilometer radius of households and markets. This distance is chosen because it is the same radius as the bandwidth for the conflict kernels, ensuring that conflict and fractionalization are measured within the same area. Fractionalization within the 50 kilometer radius around each household is used as an instrument for conflict around each household, and fractionalization within the 50 kilometer radius around each market is used as an instrument for conflict around each market. For details on how the fractionalization variable was created see box 4.3.

Ethnic Fractionalization

Because of a lack of census data for the Democratic Republic of Congo, a spatial approach was chosen to estimate ethnic fractionalization. This study uses *Peoples of Africa Atlas: An Enthnolinguistic Atlas* (Felix and Meur 2001), which was later digitized by the AfricaMap project of the Harvard Center for Geographic Analysis.

(continued next page)

Box 4.3 (continued)

This data set is used to estimate a fractionalization index similar to that of the Herfindahl Index of ethnolinguistic group shares (Alesina et al. 2003), according to equation (1):

$$FRACT_w = 1 - \sum_K p_k^2, \gamma \in 50 \text{ kilometers}, \tag{1}$$

in which p_k is the percentage of land in which ethnicity k is the dominant ethnic group, within γ (that is, the 50 kilometer bandwidth of location w), where w can be a household or a market. Higher (lower) values of the fractionalization index indicate higher (lower) levels of ethnic fractionalization. In the extreme case, if only one group inhabits a region, then $p_k = 1$, hence $FRACT_w = 0$. Specifically, this measure gives the probability that any two points of land chosen at random will have different dominant ethnic groups. For example, to estimate the fractionalization index in Kisangani, a circle or buffer of a 50 kilometer radius is created, and the area that each of the eight ethnicities occupies within that circle is estimated to arrive at the percentage of land area that each tribe or ethnicity dominates in the areas surrounding Kisangani. Those percentages are plugged into equation (1) to arrive at the fractionalization index value of 0.83, which means that Kisangani is an area with a very high level of ethnic fractionalization (see figure B4.3.1). For more information on the fractionalization index, see appendix A.

Figure B4.3.1 Ethnic Fractionalization around Kisangani

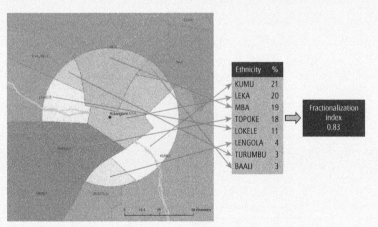

Source: Calculated from Felix and Meur (2001).

Estimating Impacts of Roads and Conflict and Their Combined Effect on Welfare

In estimating the effects of conflict and transport costs on the well-being of households, two alternative specifications are considered. The first test examines whether the impact of conflict differs depending on whether it occurs near markets or near households. Then, the combined effect of both transport costs and conflict on household well-being is examined by including an interactive term. Box 4.4 discusses each of these specifications in greater detail and explains the strategy used to estimate the effects.

BOX 4.4

Estimation Technique

Conflict Near Market versus Conflict Near Household

Presumably, the effects of conflict near markets could be avoided by retreating to subsistence modes of livelihood, whereas conflict near the home is likely to be more devastating because its negative effects cannot be avoided, even by adopting subsistence farming. This study uses the following specification to investigate whether the impacts of conflict differ depending on where it is located relative to the individual household:

$$\ln(W_i) = \beta_0 + \beta_1 \ln(T_i) + \beta_M \ln(C_i^M) + \beta_H \ln(C_i^H) + X_i'\gamma + \varepsilon_i. \tag{1}$$

In equation (1), W_i represents the well-being of the household, measured by the wealth index, the multidimensional poverty index, or local GDP. T_i represents transport cost to the nearest market and is instrumented using the natural-historical path variable. Conflict, measured by the kernelly estimated number of fatalities within a 50 kilometer radius of the market and household, is given by C_i^M and C_i^H, respectively. These conflict terms are instrumented using the fractionalization index, its squared term (to account for the diminishing marginal effects of fractionalization on a fixed portion of land), and the Euclidean distance to the eastern border, given that there is greater conflict along this border and because distance to this border should not directly affect income. When local GDP is the dependent variable, the location of the household is simply replaced with the location of the centroid of

(continued next page)

Box 4.4 (continued)

the grid cell. The coefficients of these variables provide an indication of the differential effects of transport costs.

Finally, X_i denotes a vector of control variables that are likely to affect household well-being. Given the difference between household and spatial data, different control variables are used in the regressions using the two data sets. For the household data, this vector includes agricultural variables, including agricultural potential (because areas with greater agricultural potential may naturally be more likely to have greater wealth) and a dummy indicating whether the household is engaged in agricultural activities (because agricultural households may accumulate wealth differently from others). Household demographic characteristics are also controlled for, including the age of the household head, a binary variable indicating whether the household head is female, number of female members age 15–49, number of male members age 15–59, and number of children age 0–5 (with all continuous variables estimated in log form). These variables help account for the fact that households with different demographic characteristics will have different propensities to accumulate wealth, as well as different levels of health and education. Finally, the regression includes a binary variable indicating whether the household is in a rural area and fixed effects indicating the geographic zone (to account for the unobserved characteristics of the area in which the household resides). For local GDP, controls include a quadratic term of population within the grid cell (to control for agglomeration benefits), the agro-ecological potential yield of several important crops within the grid cell, the distance to the nearest mining facility, and province fixed effects. All continuous variables are included in log form.

Joint Effects of Conflict and Transport Costs

To examine interactions between conflict and transport costs, a binary variable for high-conflict areas is created and introduced into the model as an interactive term. Specifically, equation (1) is modified as follows:

$$\ln(W_i) = \beta_0 + \beta_1 \ln(T_i) + \beta_2 dc_i + \beta_3 \ln(T_i) \times dC_i + X_i' \gamma + \varepsilon_i. \qquad (2)$$

In equation (2), the binary variable dC_i takes the value of 1 if conflict near the household and markets are both greater than the median, and zero otherwise. The other variables are defined as in equation (1). By including an interaction term ($\ln(T_i) \times dC_i$), it is possible to estimate how transport costs affect welfare and economic activity in areas with a high level of conflict.

(continued next page)

Box 4.4 (continued)

As before, the variable indicating the time taken to reach the nearest market (the natural-historical path variable) is used to instrument for transport cost to market, and social fractionalization (in quadratic form) and distance to eastern border are used to instrument for conflict. All of these instruments are in log form.

As a robustness check, a set of Conley bounds are calculated for the coefficients of interest (see chapter 2, box 2.2). Equation (3) illustrates how the Conley bounds are calculated:

$$\ln(W_i) = \beta_0 + \beta_T \ln(T_i) + \beta_M \ln\left(C_i^M\right) + \beta_H \ln\left(C_i^H\right) + X_i'\gamma + \lambda_0 NHP_i \tag{3}$$
$$+ \sum_{K=\{H,M\}} \lambda_{1K} F_i^K + \sum_{K=\{H,M\}} \lambda_{2K} (F_i^K)^2 + \sum_{K=\{H,M\}} \lambda_K^D D_i^K + u_i.$$

The traditional instrumental variable strategy assumes that the parameters λ_0, λ_{1k}, λ_{2k}, and λ_K^D in equation (3) are all equal to zero. The Conley bound specification, shown in Conley, Hansen, and Rossi (2012), allows these parameters to be close to zero, but not actually equal to zero. In other words, the instrumental variables are allowed to be only plausibly exogenous. By allowing the values of these parameters to vary, it is possible to test whether coefficient estimates are robust to the instrumental variables being only plausibly exogenous. (See table 4B.7 in annex 4B for the Conley bound estimates.)

Main Findings

This section presents the estimated impacts of transport costs and conflict on three outcome measures: a wealth index, the probability of being multidimensionally poor, and local GDP. Table 4.2 presents the results from regressing the wealth index (columns 1 and 2) and multidimensional poverty indicator dummy (columns 3 and 4) on transport cost and conflict near markets and households, providing estimates of equation (1) in box 4.4. Columns 1 and 2 (ordinary least squares [OLS] and two-stage least squares, respectively) both indicate that transport costs have a significantly negative effect on the wealth index. The IV result in column 2 shows that a 10 percent decrease in transport cost to market increases the wealth index by about 1.7 percent, significant at the 1 percent level. This finding indicates that

Table 4.2 **Effect of Transport Cost and Conflict on Welfare Indicators in the Democratic Republic of Congo**

	Wealth index		Multidimensional poverty indicator	
	(1)	(2)	(3)	(4)
	OLS	IV	OLS	IV
ln(transport cost to market)	−0.0903***	−0.170***	0.0222***	0.0235**
	(−9.679)	(−10.04)	(4.094)	(2.398)
ln(number of fatalities near markets in the past five years, 50 km kernel)	0.0125	0.0203	0.00348	0.0613***
	(0.826)	(0.482)	(0.376)	(2.646)
ln(number of fatalities near households in the past five years, 50 km kernel)	0.0532***	−0.359***	0.00785	0.0230
	(2.998)	(−5.163)	(0.935)	(0.693)
Other controls	Agricultural potential of area, indicator of whether household is agriculturally involved, age of household head, sex of household head, number of household members, number of males and females that are capable of working, number of children age 0–5 in the household, dummy indicating whether household is located in urban or rural area, dummy variables indicating regional location of household			
Observations	6,550	6,550	6,550	6,550

Sources: DRC Demographic and Health Survey (MPSMRM, MSP, and ICF International 2007) FAO and IIASA 2000; Raleigh et al. 2010; and calculations.
Note: IV = instrumental variable; OLS = ordinary least squares. Robust *t*-statistics in parentheses.
***p < 0.01, **p < 0.05, *p < 0.10.

households in areas with better access to markets are able to accrue more wealth than those with poor access to markets. The results also indicate that conflict near households is highly detrimental to the wealth of households— a 10 percent increase in fatalities from conflict decreases wealth by approximately 3.6 percent. Note that the effect of conflict on wealth estimated in the OLS model is a counterintuitive positive effect, as shown in column 1. This outcome indicates the OLS estimation is subject to significant bias, and it is necessary to use the IV estimation technique.

Columns 3 and 4 of table 4.2 report a significantly positive impact of transport costs on the probability of being multidimensionally poor. The IV results in column 4 show that a 10 percent decrease in transport costs decreases the probability of being multidimensionally poor by about 0.24 percent. Conflict around markets has a statistically significant positive effect on multidimensional poverty, indicating that a 10 percent increase in the number of fatalities around markets increases the probability of being multidimensionally poor by approximately 0.6 percent. Conflict near households has no statistical effect on the multidimensional poverty index. These results suggest that conflict near households matters more than conflict near markets for the wealth index relative to the probability of being multidimensionally poor. This result may occur because conflict near households allows rebels easy access to household wealth. The results related to multidimensional poverty indicate that conflict near markets matters more than conflict near households, possibly because educational institutions and health facilities are located near markets, so conflict near markets hampers access to these facilities and thereby affects these components of the multidimensional poverty index more directly.

Next, the combined (interactive) effect of transport cost on the wealth index and multidimensional poverty indicator is examined. These results are shown in table 4.3. Column 2 of table 4.3 shows that when conflict near both the household and the closest market is low, the effect of transport costs on the wealth index is negative and statistically significant. In this low-conflict scenario, a 10 percent decrease in transport costs increases the wealth index by 1.14 percent. When conflict near both the household and the closest market is high, the effect

Table 4.3 Combined Effect of Transport Cost and Conflict on Wealth Index and Multidimensional Poverty

	Wealth index		Multidimensional poverty indicator	
	(1)	(2)	(3)	(4)
	OLS	IV	OLS	IV
ln(transport cost to market)	−0.0865***	−0.114***	0.0199***	0.0224***
	(0.00936)	(0.0134)	(0.00560)	(0.00719)
Dummy: Conflict near household and markets high	0.170***	−0.00177	−0.0167	−0.00709
	(0.0356)	(0.102)	(0.0262)	(0.0484)
High-conflict dummy interacted with log of transport cost	−0.00812	0.132***	−0.000246	−0.00571
	(0.0139)	(0.0327)	(0.0103)	(0.0185)
Other controls	Agricultural potential of area, indicator of whether household is agriculturally involved, age of household head, sex of household head, number of household members, number of males and females that are capable of working, number of children age 0–5 years in the household, dummy indicating whether household is located in urban or rural area, and dummy variables indicating regional location of household			
Observations	6,550	6,550	6,550	6,550
Hansen J statistic (overidentification test of all instruments):		155.757		90.963
Chi-sq(10) P-value		0.0000		0.0000

Sources: DRC Demographic and Health Survey (MPSMRM, MSP, and ICF International 2007); FAO and IIASA 2000; Raleigh et al. 2010; and calculations.
Note: IV = instrumental variable; OLS = ordinary least squares. Variables used to instrument for dummy indicating that conflict near both household and market are above the median level are distances to eastern border from household and market, levels of social fractionalization and its squared term near household and market. Robust t-statistics in parentheses.
***$p < 0.01$, **$p < 0.05$, *$p < 0.10$.

of transport costs is given by the sum of the coefficients on transport costs and the interaction term. In this high-conflict scenario, a 10 percent decrease in transport costs decreases the wealth index by 0.18 percent. This result indicates that when conflict is high near both the closest market and the household, households that are farther away from markets are probably better off. This scenario is likely because lack of access to markets provides a measure of protection from conflict. This is a new finding in the literature, which suggests that in areas of high conflict, being near a road or a market may be detrimental to economic well-being.

The results in table 4.4 show the effect of reducing transport cost and conflict on economic activity measured using local GDP as the dependent variable. The IV estimate of transport cost elasticity indicates that when transport cost decreases by 10 percent, local GDP increases by 1 percent. This estimate is very close to the effect on wealth that was estimated using DHS data for the DRC.[10] Additionally, the IV estimates show that a 10 percent increase in the number of fatalities will lead to a 1.2 percent decrease in local GDP. The first-stage results indicate that the instruments strongly predict transport cost and fatalities in the area (see table 4B.3B in annex 4B).

The final set of results, in table 4.5, seeks to identify whether impacts on local GDP differ by intensity of conflict, as is suggested by the results from the DHS data in table 4.3. The regression includes a dummy variable for cases in which the number of fatalities is higher than the mean level of fatalities and an interaction term between this dummy and transport costs. The results suggest that when conflict is relatively low, a 10 percent decrease in transport cost would

Table 4.4 Effect of Transport Cost and Conflict on Local GDP

Variables	(1) OLS light, full sample	(2) IV light, full sample
ln(transport cost to market)	−0.0115**	−0.105***
	(−2.429)	(−5.425)
ln(number of fatalities near the grid cell in the past five years, 50 km kernel)[a]	0.00789***	−0.121***
	(4.761)	(−4.464)
Other controls	Population; distance to mine; agricultural potential of cassava, maize, and groundnut; province dummies	
Observations	26,330	26,330

Sources: Ghosh et al. 2010; LandScan 2006; FAO and IIASA 2000; Raleigh et al. 2010; and calculations.
Note: IV = instrumental variable; OLS = ordinary least squares. "Light" is a luminosity measure, an indicator or proxy for local GDP. Robust *t*-statistics in parentheses.
a. Only one measure of conflict is included in this specification, which is conflict near the local GDP cell. The reason for this is because, unlike the DHS data, the local GDP data set is gridded, and contains observations both in rural areas and at the markets. Given the granularity of the local GDP cells, the effect of conflict at the market will be captured in the only one or two observations that occur near each of the market locations.
***$p < 0.01$, **$p < 0.05$, *$p < 0.10$.

Table 4.5 Combined Effect of Transport Cost and Conflict on Local GDP

Variables	(1) OLS light, full sample	(2) IV light, full sample	(3) OLS light, rural only	(4) IV light, rural only
Log(transport cost)	−0.00253	−0.104***	−0.00328	−0.0974***
	(−0.495)	(−8.984)	(−0.639)	(−8.062)
Dummy: Conflict is high	0.212***	−1.251***	0.204***	−1.248***
	(6.188)	(−7.353)	(5.700)	(−6.540)
Log(transport cost) × Dummy conflict high	−0.0540***	0.324***	−0.0524***	0.334***
	(−5.326)	(4.392)	(−4.965)	(4.297)
Other controls	Population; distance to mine; agricultural potential of cassava, maize, and groundnut; province dummies			
IV for transport cost		Natural-historical path		Natural-historical path
IV for conflict		Social fractionalization and distance to DRC		Social fractionalization and distance to DRC
Observations	26,330	26,330	26,179	26,179
R^2	0.920	0.914	0.916	0.911

Sources: Ghosh et al. 2010; LandScan 2006; FAO and IIASA 2000; Raleigh et al. 2010; and calculations.
Note: IV = instrumental variable; OLS = ordinary least squares. "Light" is a luminosity measure, an indicator or proxy for local GDP. Robust *t*-statistics in parenthéses.
***$p < 0.01$, **$p < 0.05$, *$p < 0.10$.

increase local GDP by 1 percent. However, when conflict is relatively high, a 10 percent reduction in transport cost decreases local GDP by 2.2 percent. These results add further support for the DHS results in table 4.3, and indicate that when conflict is high, reducing transport cost may not be a priority intervention, perhaps because, as the model suggests, in these cases better roads enhance the payoffs from rebellion more than those from farming. When transport costs are at the mean (US$45.1) a 10 percent increase in conflict leads to a 0.8 percent decrease in local GDP. Overall, the results from the analysis of DHS data and local GDP data predict qualitatively and quantitatively consistent results.

To test the validity of the above results, robustness checks are presented that estimate a plausible range of benefits using Conley bounds. Conley bounds give a range of estimates under the assumption that the IVs employed partially violate the exclusion restriction. The estimated Conley bounds demonstrate that under the assumption that the IVs employed are improper, the estimates that are obtained all remain consistent, that is, the direction and magnitude of the estimated effects are robust. (See table 4B.7 in annex 4B for the Conley bound estimates.)

Concluding Comments

This chapter presents new results on the effects of transport infrastructure on well-being in areas of high conflict. It is widely assumed in policy circles, as indicated by the large investments, that rebuilding infrastructure that has been damaged by prolonged conflict, or perhaps even constructing new infrastructure, will invariably produce economic benefits. The results of this analysis, however, present a more nuanced story. This chapter shows that in areas with high conflict, the economic and social benefits of roads may be negated, and in some cases, actually reversed. Although these results in the context of transport infrastructure are new, they are reminiscent of past research on the effects of foreign aid and conflict. Collier and Hoeffler (2004) show that although countries tend to receive a large influx of foreign aid in the years right after a civil war has formally ended, the assistance provides little benefit during this period because of limited absorptive capacity and risk of economic bottlenecks. Instead the benefits of foreign aid are greatest after a lapse of four to seven years after the end of a civil war.

More generally, however, in areas of the DRC with low or no conflict, investment in decreasing transport cost emerges as a highly effective way to generate economic growth. Although how current or future conflicts in the DRC will evolve is unknown, this study has shown that upon the cessation of conflict, new road construction in one of the most isolated and disconnected countries in the world can be an important tool for catalyzing growth and ending the perpetual conflict trap. Therefore, the results in this chapter do not call for an unqualified proscription of road investments in all conflict-prone areas. At a minimum, however, they do suggest the need to be mindful of unintended consequences and the desirability of combining such investments with a commitment to enhanced security to realize the benefits.

Annex 4A Theoretical Framework

To guide the empirical analysis a simple theoretical model is outlined that describes the manner in which variations in access to markets might influence incentives for rebellion in conflict-affected areas. In the model, mobilization decisions are endogenous and the focus is on the effects of transport costs on welfare, production, and the conflict incentives of individuals. Although a significant theoretical literature on the economic determinants of conflict is available, to the authors' knowledge none of the models explore the role of transport infrastructure on rebel incentives.[11] These issues are arguably important for understanding how policies influence individual decisions to engage in productive activities or to join rebel forces, and for guiding the design of development strategies.

The model analyzes a situation in which households at spatially distinct locations have a choice between either joining a rebel group that loots from other households or engaging in some productive activity, such as farming. The role of the government is left in the background, with levels of enforcement taken as given. Although not every incidence of civil conflict is of this form, the DRC and many others fit into this broad category in which government control is limited and violence often takes the form of looting rather than an outright struggle against the authorities whose power and presence is often circumscribed (see Fearon [2007] for a discussion).

In this framework, three mechanisms are key to understanding when insurgency is rendered more (or less) profitable than some other economic activity. The first is the usual opportunity cost of conflict: when incomes are higher (say, because of lower transport costs), the forgone income from joining the insurgents will be greater, so there is less incentive to participate in rebellion. The second mechanism concerns the size of the prize that is available to the rebels. In most theoretical models the prize is exogenously given and is typically defined by the availability of natural resources or by exogenous factors such as rainfall (Miguel, Satyanath, and Sergenti 2004). In the current context the amount that can be looted is endogenously determined by the productive activities of households. Because rebels loot from farm households, higher levels of farmer incomes increase the lootable prize and make conflict more attractive. The third mechanism operates as follows: as payoffs to productive economic activities such as farming decline, the incentive to join the rebels becomes greater. All else equal, a smaller lootable prize will need to be shared among a larger number of looters. The results, therefore, suggest that there is no simple linear relationship between policies that promote access to markets and development outcomes. Circumstances in which such strategies can both inflame and moderate conflict are identified. These issues appear not to have been explored and formally modeled or empirically tested in the literature in this context.

The simplest possible economic structure and functional forms are considered to generate empirically testable solutions. The economy consists of a continuum of individuals who are uniformly distributed along the unit interval $n_i \in [0,1]$, where the individual at location $n_i = 0$ is closest to the market and that at $n_i = 1$ is farthest away from the market. Transport costs to the market are given by $t_i = zn_i$, $z > 0$. Individuals at each location can either choose to farm or join a rebel force that loots from those who farm.

Decisions are made sequentially. In the first stage each individual decides whether to join the rebels or farm. Given the set of rebels (or equivalently, the set of farmers at known locations), in the second stage the rebels determine effort levels and a looting strategy that consists of the decision of whether to attack at the market or where the farmers are located. In the third stage farmers determine the types of goods to produce and the production levels of each good. There are two types of produced goods—those produced for sale in the markets, which must incur transport costs, and goods for domestic consumption (subsistence farming). Both products are potentially vulnerable to theft by rebels. This structure implies that the rebels have a strategic early-mover advantage relative to farmers. The model is solved by backward induction, beginning with the final stage of the game.

Stage 3—The Farmers' Decisions

In the third stage farmers determine production decisions. Each farmer is endowed with L units of an input that can be used to produce either a marketed good denoted L_m or a domestically consumed subsistence good denoted L_c, where L is the fixed endowment with $L = L_m + L_c$. The production functions for the two products are given by $M = mL_m$ and $C = cL_c$ (where $m > 0$ and $c > 0$). Goods produced for the market are sold to purchase a composite commodity denoted B, which is consumed. Each farmer determines production levels to maximize utility:

$$U_F = B^\beta \left(C(1-f)\right)^\alpha, \tag{4.1}$$

in which C is the self-consumed product produced by the farmer and f is the fraction of this product that is seized by rebels at the location of the farmer; $\alpha > 0$, $\beta > 0$ with $\alpha + \beta = 1$.[12] Thus, equation (4.1) defines the utility to the farmer net of theft. Utility is maximized subject to the budget constraint:

$$(t_i + P_B)B = (1 - g)M(P - t_i), \tag{4.2}$$

in which $t_i = zn_i$ is the cost of transport, P_B = the price of purchased good B, P is the price of the marketed (sold) good, and g is the proportion of the marketed good stolen by the rebels at the market.[13] Equation (4.2) simply asserts that the amount of money spent on purchased goods, including transport costs $(t_i + P_B)$, must equal the amount received from selling goods at the market net of transport costs $(P - t_i)$. A fraction g of these marketed goods is stolen by rebels.

Maximizing equations (4.1) and (4.2) with respect to B, L_m, and L_c generates the following reaction functions:

$$B = \frac{\beta(1-g)(P-t_i)\beta mL}{(P_B + t_i)}, \tag{4.3}$$

$$L_m = \beta L, \tag{4.4}$$

$$L_c = \alpha L. \tag{4.5}$$

Substituting in equation (4.1) yields the indirect utility function for this stage:

$$U_F^*(t_i) = \left[\frac{\beta(1-g)(P-t_i)\beta mL}{(P_B+t_i)}\right]^\beta [c\alpha L(1-f)]^\alpha. \tag{4.6}$$

Note for future reference that $\dfrac{dU_F(t_i)}{dt_i} < 0$. Unsurprisingly in this simple set up, higher transport costs unambiguously lower welfare.

Stage 2—The Rebel's Problem

Turning next to the rebel's problem, in stage 2 rebels determine their strategy, which consists of deciding whether to loot from the market or from the farm. For simplicity it is assumed that all looted goods are valued equally by the rebels, implying that the results are not influenced by arbitrary assumptions about relative prices of the goods. In this stage of the game the set of farmers and rebels is taken as given. Let n^* be the given set of farmers (determined in stage 1); then aggregate payoffs to the rebel group is given by the following expression:

$$U_R = \int_0^{n^*} \left(gmL_m + fc(L-L_m)\right)dx - K_g g^2 - K_f f^2, \tag{4.7}$$

in which K_g, K_f are costs of looting at the market and farm, respectively.[14] It is assumed that these costs include the risks and consequences of resisting the given levels of government defense. It can be demonstrated that the existence of an interior equilibrium with both farmers and rebels is contingent upon these costs being at intermediate levels. Excessively high looting costs render rebellion unattractive and vice versa. The rebels maximize equation (4.7) taking as given the stage 3 decisions of farmers. Thus, substituting from equation (4.4) for L_m and maximizing with respect to g and f yields the rebels' aggregate distribution of looting between farms and households:

$$\hat{g} = \frac{n^* m\beta L}{2K_g} \tag{4.8}$$

$$\hat{f} = \frac{n^* c(1-\beta)L}{2K_f}. \tag{4.9}$$

Substituting from equations (4.8) and (4.9) yields the rebel group's aggregate indirect payoff function:

$$U_R^* = \frac{(n^*)^2}{4}\left(\left\{\frac{\left[cL(1-\beta)\right]^2}{K_f}\right\} + \left\{\frac{\left[mL\beta\right]^2}{K_g}\right\}\right). \tag{4.10}$$

It is assumed that these benefits are distributed equally between rebels, so that payoffs to the individual rebel j is simply

$$U_{Rj}^* = \frac{U_R^*}{(1-n^*)}. \tag{4.11}$$

Stage 1—The Decision to Farm or Rebel

In the first stage of the game each agent decides whether to farm or join the rebels, given knowledge of the downstream responses. To simplify the analysis, the problems associated with imperfect or costly monitoring and shirking that might occur in the rebel group are removed. The simplest possible case is taken and it is assumed that the marginal agent switches from farming to join the rebels if the payoffs from farming are less than those from joining the rebels. Hence, the marginal farmer, located at $\widehat{n}_i$, is indifferent between farming and joining the rebels if

$$U_F\left(\widehat{n}_i\right) = U_R\left(\widehat{n}_i\right). \tag{4.12}$$

The value $\widehat{n}_i$ is the solution to

$$\psi \equiv U_F\left(\widehat{n}_i\right) - U_R\left(\widehat{n}_i\right) = 0. \tag{4.13}$$

Substituting equations (4.8) and (4.9) into equation (4.13) yields

$$\begin{aligned}\psi \equiv{} & \left[\frac{\beta(1-g)(P-t_i)\beta mL}{(P_B+t_i)}\right]^\beta \left[c\alpha L(1-f)\right]^\alpha \\[2mm]
& - \frac{n^2}{4(1-n)}\left(\left\{\frac{\left[cL(1-\beta)\right]^2}{K_f}\right\} + \left\{\frac{\left[mL\beta\right]^2}{K_g}\right\}\right).\end{aligned} \tag{4.14}$$

It is useful to explore how changes in transport costs influence the decision to join rebel forces. In general, higher transport costs are likely to have ambiguous effects. Intuitively, higher transport costs lower output levels and the utility from farming, which makes rebellion more attractive, all else equal. However, lower output levels also reduce the amount that is available for looting, so the benefits of looting decline too. In addition, if some households switch from farming to looting, a smaller amount of output will need to be shared among a larger number of looters. The overall decision to farm (or participate in rebellion) will then depend on the relative rates of decline in the payoffs from farming and looting. If farming utility falls more rapidly than that from rebellion, the marginal agent will switch from farming to the rebel forces and vice versa. Result 1 summarizes the cases when higher (lower) transport costs induce less (more) conflict.

Result 1: Lower transport costs will induce a switch from farming to rebels (and vice versa) when the cost of stealing marketed goods is sufficiently low (high) (that is, $\lim_{K_g \to 0}$, *then* $\dfrac{d\psi}{dt} \to -\infty$).

The proof appears later in this section.

Intuitively, lower transport costs make production for the market more attractive. Recall that K_g defines the cost of looting. As the costs of looting from the market decline, the strategic advantage (that is, payoffs) accruing to the rebels increases because they are able to loot a greater share of the aggregate output produced by farmers. As a result, payoffs available to the rebels from theft of the higher aggregate output can rise faster than the increased output of the marginal farmer. A corollary of Result 1 is that when more households join the rebels, agricultural output would decline in such situations, all else equal, so there can be no presumption that transport cost reductions induce the desired outcomes.

The next result explores whether an increase in conflict at the market is more damaging than at the household.[15]

Result 2: In general, the welfare cost of a marginal increase in looting of subsistence goods relative to marketed goods is ambiguous, and is increasing in the relative welfare weights of these goods in the utility function (that is,
$$\left| \frac{dU_i^F}{df} \right| > \left| \frac{dU_i^F}{dg} \right| if \frac{\alpha}{\beta} > \frac{1-f}{1-g}).$$

The proof appears later in this section.

Result 2 suggests that the consequences of increased looting at each location depend critically upon the welfare weights in the utility function. When the marketed good is given sufficient weight relative to the subsistence product, the marginal welfare costs of theft at the market are higher and vice versa. An implication of this result is that the combined effects of simultaneous changes in transport costs and conflict are also ambiguous even in this highly stylized framework.

In sum, the model shows that transport costs affect the consumption and utility levels of households through numerous channels, one by directly affecting the revenue of the households, and others indirectly by affecting the amount available to loot and the number of looters. The model also suggests that lower transport costs could induce more conflict, and ultimately lead to a reduction in welfare in conflict-prone areas (Result 1). These empirically testable hypotheses are tested using data from the DRC. In addition, the effects of conflict at markets and households are empirically analyzed (Result 2), as is the combined effect of conflict and transport cost, given that the theory is ambiguous in predicting the consequences.

Proof of Results from Theoretical Model
Heuristic Proof for Existence of the Equilibrium
Without loss of generality, assume that $K_f = K_g = K$. Observe that as $K \to 0$ then

$$U_R(n) = \frac{n^2}{4(1-n)} \left(\left\{ \frac{\left[cL(1-\beta)\right]^2}{K} \right\} + \left\{ \frac{\left[mL\beta\right]^2}{K} \right\} \right) \to \infty \text{ and}$$

$$U_F(n_i) = \left[\frac{\beta(1-g)(P-t_i)\beta mL}{(P_B+t_i)} \right]^\beta \left[c\alpha L(1-f) \right]^\alpha \to 0 \text{ (since } g \to 1 \text{ and } f \to 1).$$

Hence, as $K \to 0$ then $\psi \equiv U_F(n_i) - U_R(n_i) \to -\infty$.

Similarly, as $K \to \infty$ then $U_R(n) \to 0$ and $U_F(n_i) \to \left[\frac{\beta(P-t_i)\beta mL}{(P_B+t_i)} \right]^\beta \left[c\alpha L \right]^\alpha > 0$,

hence $\Psi > 0$. Since Ψ is continuous in K it follows that there exists some point such that $\psi \equiv U_F(n_i) - U_R(n_i) = 0$.

Result 1
Consider a change in transport costs z for the marginal household. Note that

$$\frac{d\psi}{dz} = \frac{\partial \psi}{\partial z} + \frac{\partial \psi}{\partial n}\frac{\partial n}{\partial z} \text{ and } \frac{\partial U^F}{\partial t} = \frac{\partial t}{\partial z}\left(\frac{nA^\alpha(1-g)mL\beta B^{\beta-1}(-\beta-(p-t))}{P_B+t} \right) < 0;$$

where $A = c\alpha L(1-f)$ and $B = \dfrac{\beta(1-g)(P-t_i)\beta mL}{(P_B+t_i)}$. Similarly

$$\frac{\partial U^R}{\partial t} = \frac{2-n}{4(1-n)^2}\left(\frac{\left(cL(1-\beta)\right)^2}{K_f} + \frac{(mL\beta)^2}{K_g}\right)\frac{\partial n}{\partial t}\frac{\partial t}{\partial z} < 0 \text{ (since } \frac{\partial n}{\partial z}\frac{\partial t}{\partial z} < 0 \text{ and}$$

$1 - n > 0$ and $1 - \beta > 0$ by assumption). Further note from equation (4.8) as

$K_g \to 0$ then $g \to 1$, hence $\dfrac{\partial U^F}{\partial t} \to 0; \dfrac{\partial U_R}{\partial t} \to -\infty$ and $\dfrac{\partial \psi}{\partial t} \to \infty.$[16]

Result 2

$$\frac{\partial U^F}{\partial f} = -c(1-\beta)B^\beta L\alpha A^{\alpha-1} \text{ and } \frac{\partial U^F}{\partial g} = \frac{-\beta B^{\beta-1} LA^\alpha (p-t)}{(p_B+t)}. \text{ Thus}$$

$$\left|\frac{\partial U^F}{\partial f}\right| - \left|\frac{\partial U^F}{\partial g}\right| > 0 \text{ whenever}$$

$$\frac{LA^{\alpha-1}B^{\beta-1}}{p_B+t}(p-t)(1-\beta)cmL\beta\big(\alpha(1-g)-\beta(1-f)\big) > 0, \text{ which holds whenever}$$

$$\frac{\alpha}{\beta} > \frac{1-f}{1-g}.$$

Annex 4B Summary Statistics and Regression Results

Table 4B.1 Summary Statistics from Demographic and Health Survey

Variable	Mean	Standard deviation	Minimum	Maximum
Outcomes				
Wealth index	4,621.605	103,646.1	−107,521	345,191
Multidimensional poverty indicator (dummy)	0.817	0.387	0	1
Variables of interest				
Transport cost to market	19.834	21.695	0.734	124.253
Number of fatalities around the nearest market in the past five years	0.044	0.144	0	0.907
Number of fatalities around household in the past five years	0.017	0.073	0	0.815
Dummy = 1 if conflict near household and market is high	0.328	0.470	0	1

(continued next page)

Table 4B.1 (continued)

Variable	Mean	Standard deviation	Minimum	Maximum
Instruments				
Time taken to reach market using natural and historical path	23.559	26.485	0.803	152.413
Fractionalization index within 50 km radius of market	0.611	0.207	0	0.839
Fractionalization index within 50 km radius of household	0.574	0.207	0	0.846
Distance to eastern border from household (km)	880.962	486.310	0.601	1,715.954
Distance to eastern border from market (km)	882.179	486.532	2.127	1682.461
Controls				
Agricultural potential (factor of ln agricultural potential for cassava, maize, and groundnut)	−0.010	1.007	−7.475	1.851
Dummy: Household agriculturally involved	0.543	0.498	0	1
Age of household head	41.473	13.032	15	93
Dummy: Female household head	0.189	0.391	0	1
Number of household members	5.846	2.961	1	28
Number of female household members age 15–49 years	1.345	0.855	0	9
Number of male household members age 15–59 years	0.658	0.941	0	13
Number of children in household age 0–5 years	0.582	0.936	0	10
Dummy = 1 if type of residence is rural	0.555	0.497	0	1
Dummy = 1 if region is Bas-Congo	0.095	0.294	0	1
Dummy = 1 if region is Bandundu	0.107	0.309	0	1
Dummy = 1 if region is Equateur	0.104	0.305	0	1
Dummy = 1 if region is Orientale	0.077	0.267	0	1
Dummy = 1 if region is Nord-Kivu	0.032	0.177	0	1
Dummy = 1 if region is Maniema	0.094	0.291	0	1
Dummy = 1 if region is Sud-Kivu	0.043	0.202	0	1
Dummy = 1 if region is Katanga	0.109	0.311	0	1
Dummy = 1 if region is Kasaï-Oriental	0.103	0.304	0	1
Dummy = 1 if region is Kasaï-Occidental	0.092	0.289	0	1
Number of observations: 6,728				

Sources: DRC Demographic and Health Survey (MPSMRM, MSP, and ICF International 2007); FAO and IIASA 2000; Raleigh et al. 2010; and calculations.

Table 4B.2 Summary Statistics, Local GDP

Variable	Mean	Standard deviation	Minimum	Maximum
Local GDP/total income per cell (million US$, 2006 PPP)	0.836	23.577	0.836	23.577
Transport cost to market (US$)	45.103	28.444	45.103	28.444
Number of fatalities within 50 km of cell (kernelly estimated)	0.006	0.027	0.006	0.027
Population per cell (thousands), LandScan 2006	2,517.983	28,082.600	2,517.983	28,082.600
Cassava yield (kg/ha; GAEZ FAO)	1,930.541	712.775	1,930.541	712.775
Maize yield (kg/ha; GAEZ FAO)	949.325	362.901	949.325	362.901
Groundnut yield (kg/ha; GAEZ FAO)	184.517	124.652	184.517	124.652
Distance to nearest mine (km)	337.376	147.466	337.376	147.466
Time to reach market using natural-historical path (hours)	48.372	32.275	48.372	32.275
Fractionalization index within 50 km radius of cell	52.729	21.050	52.729	21.050
Distance from cell to market (km)	682,277.300	363,717.000	682,277.300	363,717.000
Conflict near cell is high (dummy)	0.110	0.312	0.110	0.312
Number of observations = 26,330				

Sources: Ghosh et al. 2010; LandScan 2006; FAO and IIASA 2000; Raleigh et al. 2010; and calculations.
Note: GAEZ FAO = Global Agro-Ecological Zones from FAO and IIASA (2000); kg/ha = kilograms per hectare; PPP = purchasing power parity.

Table 4B.3A Effect of Transport Cost and Conflict on Welfare Indicators in the Democratic Republic of Congo

	Wealth index		Multidimensional poverty indicator	
	(1)	(2)	(3)	(4)
Variables	OLS	IV NHP	OLS	IV NHP
Log(transport cost)	−0.0903***	−0.170***	0.0222***	0.0235**
	(−9.679)	(−10.04)	(4.094)	(2.398)
Log(number of fatalities near markets in the past five years) (50 km kernel)	0.0125	0.0203	0.00348	0.0613***
	(0.826)	(0.482)	(0.376)	(2.646)
Log(number of fatalities near households in the past five years) (50 km kernel)	0.0532***	−0.359***	0.00785	0.0230
	(2.998)	(−5.163)	(0.935)	(0.693)
Agricultural potential	0.00435	−0.00634	−0.00201	−0.00175
	(0.395)	(−0.535)	(−0.346)	(−0.301)
Dummy: Household agriculturally involved	−0.384***	−0.437***	0.0898***	0.0980***
	(−16.69)	(−15.91)	(6.810)	(7.209)
Log(age of household head)	−0.0253	−0.0268	−0.0515***	−0.0508***
	(−0.859)	(−0.791)	(−2.870)	(−2.803)
Dummy: Female household head	−0.139***	−0.147***	0.0531***	0.0523***
	(−5.880)	(−5.813)	(3.761)	(3.675)
Log(number of household members)	0.103***	0.0806***	0.0603***	0.0645***
	(3.763)	(2.674)	(3.491)	(3.716)

(continued next page)

Table 4B.3A (continued)

Variables	Wealth index		Multidimensional poverty indicator	
	(1)	(2)	(3)	(4)
	OLS	IV NHP	OLS	IV NHP
Log(number of female household members age 15–49 years)	0.188***	0.207***	−0.0793***	−0.0825***
	(5.807)	(5.804)	(−4.136)	(−4.283)
Log(number of male household members age 15–59 years)	0.130***	0.141***	0.0308**	0.0313**
	(6.288)	(6.336)	(2.292)	(2.322)
Log(number of children in household age 0–5 years)	−0.109***	−0.105***	0.0312**	0.0290**
	(−5.240)	(−4.683)	(2.546)	(2.358)
Dummy: Rural area	−0.539***	−0.531***	0.0982***	0.110***
	(−19.65)	(−16.25)	(6.246)	(6.123)
Constant	12.55***	11.36***	0.571***	0.785***
	(107.7)	(48.09)	(8.586)	(8.960)
Regional fixed effects	Yes	Yes	Yes	Yes
Observations	6,550	6,550	6,550	6,550
R^2	0.654	0.599	0.251	0.243

Table 4B.3B Effect of Transport Cost and Conflict on Welfare Indicators in the Democratic Republic of Congo: First-Stage Results

	(1)	(2)	(3)
	ln(transport cost to market)	ln(number of fatalities near market)	ln(number of fatalities near household)
ln(natural-historical path)	0.977***	0.061***	−0.141***
	(300.45)	(9.95)	(−21.15)
Fractionalization near market	−0.047	−0.858***	1.292***
	(−0.51)	(−5.85)	(5.24)
Fractionalization near market^2	−0.218***	1.352***	−1.384***
	(−2.25)	(7.93)	(−5.25)
Fractionalization near household	0.2636***	1.532***	−0.708***
	(3.69)	(13.13)	(−3.7)
Fractionalization near household^2	−0.058	−1.973***	1.235***
	(−0.80)	(−15.40)	(6.09)
ln(distance from household to eastern border in km)	−0.016***	−0.279***	−0.282***
	(−2.09)	(−16.62)	(−15.86)
ln(distance from market to eastern border in km)	−0.043***	−0.476***	−0.166***
	(−3.02)	(−17.18)	(−3.52)
Angrist-Pischke test of weak identification	9559.55	179.92	121
P-value	0.0000	0.0000	0.0000

Sources: DRC Demographic and Health Survey (MPSMRM, MSP, and ICF International 2007); FAO and IIASA 2000; Raleigh et al. 2010; and calculations.
Note: IV NHP = instrumental variable natural-historical path; OLS = ordinary least squares. t-statistics in parentheses.
***$p < 0.01$, **$p < 0.05$, *$p < 0.10$.

Table 4B.4A Combined Effect of Transport Cost and Conflict on Wealth Index and Multidimensional Poverty

	Wealth index		Multidimensional poverty indicator	
	(1)	(2)	(3)	(4)
	OLS	IV	OLS	IV
Log(transport cost)	−0.0865***	−0.114***	0.0199***	0.0224***
	(−9.241)	(−8.508)	(3.550)	(3.119)
Conflict near household and markets high (dummy)	0.170***	−0.00177	−0.0167	−0.00709
	(4.772)	(−0.0173)	(−0.636)	(−0.147)
High-conflict dummy × log(transport cost)	−0.00812	0.132***	−0.000246	−0.00571
	(−0.583)	(4.028)	(−0.0239)	(−0.308)
Agricultural potential	0.0123	0.0202	−0.00289	−0.00316
	(1.069)	(1.639)	(−0.486)	(−0.487)
Household agriculturally involved (dummy)	−0.371***	−0.358***	0.0851***	0.0841***
	(−16.38)	(−14.87)	(6.404)	(6.266)
Log of age of household head	−0.0230	−0.0199	−0.0517***	−0.0518***
	(−0.798)	(−0.681)	(−2.886)	(−2.908)
Female household head (dummy)	−0.139***	−0.142***	0.0552***	0.0555***
	(−6.010)	(−6.109)	(3.985)	(4.007)
Log(number of household members)	0.0960***	0.0979***	0.0601***	0.0602***
	(3.591)	(3.630)	(3.472)	(3.490)
Log(number of female household members age 15–49 years)	0.184***	0.187***	−0.0776***	−0.0776***
	(5.763)	(5.764)	(−4.055)	(−4.047)
Log(number of male household members age 15–59 years)	0.132***	0.130***	0.0307**	0.0309**
	(6.443)	(6.315)	(2.293)	(2.308)
Log(number of children in household age 0–5 years)	−0.106***	−0.105***	0.0331***	0.0330***
	(−5.225)	(−5.133)	(2.732)	(2.727)
Rural area (dummy)	−0.535***	−0.556***	0.0948***	0.0948***
	(−20.21)	(−20.72)	(6.164)	(6.128)
Constant	12.19***	12.24***	0.551***	0.545***
	(113.0)	(85.93)	(8.238)	(7.063)
Regional fixed effects	Yes	Yes	Yes	Yes
Observations	6,728	6,728	6,728	6,728
R^2	0.657	0.651	0.248	0.248

Table 4B.4B Combined Effect of Transport Cost and Conflict on Wealth Index and
Multidimensional Poverty: First-Stage Results

	(1)	(2)	(3)
	ln(transport cost)	Dummy: Conflict near households and markets high	ln(transportation cost) × Conflict near households and markets high dummy
ln(natural-historical path)	0.952***	0.171***	0.930***
	(24.82)	(4.17)	(6.72)
Fractionalization near market	−2.768***	−0.958***	2.987***
	(−9.65)	(−1.9)	(2.44)
Fractionalization near market²	2.320***	−0.863**	−5.067***
	(8.22)	(−1.75)	(−3.82)
Fractionalization near household	3.22***	2.611***	−0.364
	(13.33)	(6.03)	(−0.37)
Fractionalization near household²	−2.67***	−0.353	1.32
	(−11.07)	(−0.83)	(1.24)
ln(distance from households to eastern border in km)	0.365***	0.0273	0.791***
	(8.6)	(0.77)	(5.85)
ln(distance from market to eastern border in km)	−0.387***	−0.091***	−0.695***
	(−7.79)	(−2.29)	(−4.43)
ln(natural-historical path) × Fractionalization near market	0.676***	0.273**	−0.638*
	(8.11)	(1.82)	(−1.73)
ln(natural-historical path) × Fractionalization near market²	−0.62***	0.212	1.230***
	(−7.84)	(1.46)	(3.09)
ln(natural-historical path) × Fractionalization near household	−0.842***	−0.826***	−0.316
	(−11.4)	(−6.68)	(−1.14)
ln(natural-historical path) × Fractionalization near household²	0.722***	0.216**	0.129
	(9.95)	(1.82)	(0.45)
ln(natural-historical path) × ln(distance from households to eastern border in km)	−0.129***	−0.006	−0.25***
	(−8.92)	(−0.52)	(−5.22)
ln(natural-historical path) × ln(distance from market to eastern border in km)	0.104***	−0.007	0.0917***
	(7.36)	(−0.6)	(1.92)
Angrist-Pischke test of weak identification	5,575.07	70.77	148.56
P-value	0.000	0.000	0.000

Sources: DRC Demographic and Health Survey (MPSMRM, MSP, and ICF International 2007); FAO and
IIASA 2000; Raleigh et al. 2010; and calculations.
Note: IV = instrumental variable; OLS = ordinary least squares. t-statistics in parentheses.
***$p < 0.01$, **$p < 0.05$, *$p < 0.10$.

Table 4B.5A Effect of Transportation Cost and Conflict on Local GDP

Variable	(1) OLS (full sample)	(2) IV (full sample)
ln(transport cost to market)	−0.0115**	−0.105***
	(−2.429)	(−5.425)
ln(number of fatalities)	0.00789***	−0.121***
	(4.761)	(−4.464)
ln(population)	0.530***	0.530***
	(170.1)	(153.3)
ln(population)^2	0.0359***	0.0357***
	(128.8)	(114.5)
ln(cassava potential yield)	0.488***	0.548***
	(5.666)	(5.648)
ln(cassava potential yield)^2	−0.0366***	−0.0395***
	(−5.774)	(−5.568)
ln(maize potential yield)	0.451***	0.348**
	(3.003)	(2.065)
ln(maize potential yield)^2	−0.0420***	−0.0388***
	(−3.686)	(−3.058)
ln(groundnut potential yield)	−0.322***	−0.384***
	(−6.244)	(−6.562)
ln(groundnut potential yield)^2	0.0422***	0.0509***
	(8.274)	(8.590)
ln(distance to mine)	−0.0764***	−0.0298***
	(−13.55)	(−2.601)
Constant	−8.953***	−9.561***
	(−42.29)	(−35.67)
Province fixed effects	Yes	Yes
R^2	0.920	0.901

Table 4B.5B **Effect of Transportation Cost and Conflict on Local GDP: First-Stage Results**

	(1)	(2)
	Full sample	
First stage results	ln(transport cost to market)	ln(number of fatalities)
ln(natural-historical path)	0.875***	−0.583***
	(356.710)	(−34.95)
Fractionalization index	−0.0003881	−0.006***
	(−1.44)	(−3.3)
(Fractionalization index)^2	0.0000136***	0.00001
	(4.490)	(0.790)
Distance to eastern border	0.0044	−0.187***
	(1.280)	(−7.97)
Angrist-Pischke test of weak identification	3,672.93	354.56
P-value	0.0000	0.0000

Sources: Ghosh et al. 2010; LandScan 2006; FAO and IIASA 2000; Raleigh et al. 2010; and calculations.
Note: IV = instrumental variable; OLS = ordinary least squares. t-statistics in parentheses.
***$p < 0.01$, **$p < 0.05$, *$p < 0.10$.

Table 4B.6A **Combined Effect of Transport Cost and Conflict on Local GDP**

	(1)	(2)
	OLS (full sample)	IV (full sample)
ln(transport cost to market)	−0.00253	−0.104***
	(−0.495)	(−8.984)
Dummy: Conflict is high = 1	0.212***	−1.251***
	(6.188)	(−7.353)
ln(transport cost to market) × Dummy: Conflict is high	−0.0540***	0.324***
	(−5.326)	(4.392)
ln(population)	0.531***	0.527***
	(170.3)	(159.2)
ln(population)^2	0.0359***	0.0358***
	(128.8)	(120.0)
ln(cassava potential yield)	0.500***	0.436***
	(5.813)	(4.830)
ln(cassava potential yield)^2	−0.0374***	−0.0329***
	(−5.901)	(−4.983)
ln(maize potential yield)	0.438***	0.489***
	(2.919)	(3.102)
ln(maize potential yield)^2	−0.0411***	−0.0462***
	(−3.611)	(−3.879)

(continued next page)

Table 4B.6A (continued)

	(1)	(2)
	OLS (full sample)	IV (full sample)
ln(groundnut potential yield)	−0.330***	−0.300***
	(−6.398)	(−5.293)
ln(groundnut potential yield)²	0.0430***	0.0413***
	(8.423)	(7.347)
ln(distance to mine)	−0.0771***	−0.0529***
	(−13.69)	(−7.205)
Province fixed effects	Yes	Yes
Observations	26,330	26,330
R^2	0.920	0.914

Table 4B.6B Combined Effect of Transport Cost and Conflict on Local GDP: First-Stage Results

First-stage results	(1) ln(transport cost to market)	(2) Conflict high (dummy)	(3) Transport cost × conflict high dummy
ln(natural-historical path)	0.985***	−0.823***	−1.067***
	(23.540)	(−17.04)	(−6.54)
Fractionalization	0.010***	−0.010***	−0.0218***
	(7.640)	(−6.68)	(−4.25)
(Fractionalization index)²	−0.0001***	0.0001***	0.0002***
	(−6.67)	(7.210)	(4.32)
Distance to eastern border	0.0185	−0.192***	−0.258***
	(1.700)	(−15.29)	(−6.06)
ln(natural-historical path) × Fractionalization	−0.003***	0.003***	0.007***
	(−8.1)	(7.13)	(4.91)
ln(natural-historical path) × (Fractionalization index)²	0.00003***	−0.00004***	−0.00008***
	(7.760)	(−8.18)	(−5.56)
ln(natural-historical path) × Distance to eastern border	−0.0042	0.0517***	0.062***
	(−1.36)	(14.63)	(5.19)
Angrist-Pischke test of weak identification	5,358.3	218.77	58.99
P-value	0.0000	0.0000	0.0000

Sources: Ghosh et al. 2010; LandScan 2006; FAO and IIASA 2000; Raleigh et al. 2010; and calculations.
Note: IV = instrumental variable; OLS = ordinary least squares. t-statistics in parentheses.
***$p < 0.01$, **$p < 0.05$, *$p < 0.10$.

Table 4B.7 Conley Bounds

		Lower bound	Upper bound
	Endogenous variable: Travel cost		
Dependent variable	IV: Natural path		
	δ: [−0.0001, 0.0001]	−0.1483	−0.1184
Wealth index	δ: [−0.001, 0.001]	−0.1493	−0.1174
	δ: [−0.01, 0.01]	−0.1588	−0.1079
	Endogenous variable: Conflict near markets		
	IV: Fractionalization near market, squared term of fractionalization near market, distance to eastern border from market		
	δ: [−0.0001, 0.0001]	−0.129	−0.0987
Wealth index	δ: [−0.001, 0.001]	−0.1292	−0.0985
	δ: [−0.01, 0.01]	−0.131	−0.0966
	IV: Fractionalization near households, squared term of fractionalization near households, distance to eastern border from households		
	δ: [−0.0001, 0.0001]	−0.2668	−0.1989
Wealth index	δ: [−0.001, 0.001]	−0.2678	−0.198
	δ: [−0.01, 0.01]	−0.2773	−0.1894
	Endogenous variable: Travel cost		
Dependent variable	IV: Natural path		
	δ: [−0.0001, 0.0001]	0.0257	0.0443
Multidimensional poverty indicator	δ: [−0.001, 0.001]	0.0247	0.0452
	δ: [−0.01, 0.01]	0.0151	0.0548
	Endogenous variable: Conflict near markets		
	IV: Fractionalization near market, squared term of fractionalization near market, distance to eastern border from market		
	δ: [−0.0001, 0.0001]	0.0164	0.0351
Multidimensional poverty indicator	δ: [−0.001, 0.001]	0.0163	0.0353
	δ: [−0.01, 0.01]	0.0144	0.0371
	IV: Fractionalization near households, squared term of fractionalization near households, distance to eastern border from households		
	δ: [−0.0001, 0.0001]	0.0192	0.0586
Multidimensional poverty indicator	δ: [−0.001, 0.001]	0.0183	0.0595
	δ: [−0.01, 0.01]	0.0093	0.0687

Sources: DRC Demographic and Health Survey (MPSMRM, MSP, and ICF International 2007; FAO and IIASA 2000; Raleigh et al. 2010; and calculations.
Note: IV = instrumental variable. Confidence intervals calculated following Conley, Hansen, and Rossi (2012).

Notes

1. Recall from chapter 2 that the "wealth index" is readily available in the Demographic and Health Survey data while the multidimensional poverty index was calculated for the purpose of this study, following Alkire and Santos (2010). Local GDP is obtained from Ghosh et al. (2010). Chapter 2 discusses in more detail the advantages of using different indicators.
2. DeLorme is a private company that specializes in global positioning system devices and has compiled a very thorough network of georeferenced roads across Africa.
3. The Highway Development Management Model is a standard model, frequently used by road engineers, that takes as inputs the road attributes available in the African Infrastructure Country Diagnostic data set, the roughness of terrain along the road, and country-level information about various factors that can affect the price of transporting goods (for example, price of fuel, labor costs, and so on) to calculate transport cost. See appendix A on Geospatial Analysis for more details.
4. The historical caravan routes were not used in previous chapters because they were unavailable for Nigeria.
5. Some recent examples of papers employing historical route IVs include Duranton, Morrow, and Turner (2014), which uses routes from major exploration expeditions in the United States between the sixteenth and nineteenth centuries as instruments for the U.S. interstate highway system; Garcia-Lopez, Holl, and Viladecans-Marsal (2013), which uses ancient Roman roads, among others, as exogenous sources of variation in Spain's current highway system; and Martincus, Carballo, and Cusolito (2012), which uses the Incan road network to instrument for Peru's current road infrastructure.
6. For a complete explanation of how this variable was created, see appendix A on Geospatial Analysis.
7. In ACLED, the level of geographic uncertainty is coded using "geoprecision codes" ranging from 1 to 3 (higher numbers indicate broader geographic spans and thus greater uncertainty about where the event occurred). A geoprecision code of 1 indicates that the coordinates mark the exact location where the event took place. When a specific location is not provided, ACLED selects the provincial capital. This way ACLED may attribute violent incidents to towns when, in fact, they took place in rural areas; this method introduces a systematic bias toward attributes associated with urban areas that can lead to invalid inferences.
8. For the analysis of local GDP, the mean number of fatalities is used instead of the median number, because the median number of fatalities in that data set is zero.
9. Using a similar argument, Mauro (1995) uses ethnolinguistic fractionalization to instrument for institutional quality when estimating its effect on GDP growth.
10. The OLS estimation results in a counterintuitive positive effect of conflict on local GDP while the IV estimate of the effect of conflict on local GDP is negative, providing further evidence that it is necessary to instrument for conflict.
11. See, for example, Bueno de Mesquita (2013) and Laitin and Shapiro (2008) and the references therein.

12. Note that the results go through the case in which $U_F = B^\beta (C)^\alpha(1-f)$ implying theft of both goods.

13. When not required, the location subscript (i) is ignored for notational brevity.

14. Though desirable for theoretical completeness, since the focus of this paper is on the empirical analysis, to save space results of the perverse case where farmers locate over the interval $[n^*,1]$ are not shown. Results are available upon request.

15. As is conventional, rebel utility is ignored.

16. Alternatively, this result may be derived as follows: $\dfrac{d\psi}{dz} = \dfrac{\partial(U^F - U^R)}{\partial z} + \dfrac{\partial(U^F - U^R)}{\partial n}\dfrac{dn}{dz}$.

From the stage 1 equilibrium condition, it is given that in equilibrium by total differentiation of Ψ one gets $\dfrac{dn}{dz} = \dfrac{\partial U^R/\partial n}{\partial U^F/\partial t}$. Substitution in the preceding equation

and rearranging yields $\dfrac{d\psi}{dz} = \left(\dfrac{\partial U^F}{\partial z}\right)^2 - \left(\dfrac{\partial U^R}{\partial n}\right)^2$. Lim $K_g \to 0$ then $g \to 1$, hence

$\dfrac{\partial U^F}{dz} \to 0; \dfrac{\partial U^R}{dn} \to -\infty$ and $\dfrac{d\psi}{dz} \to \infty$.

References

African Development Bank. 2003. "Review of the Implementation Status of the Trans African Highways and the Missing Links, Volume 1: Main Report." SWECO International, AB, and Nordic Consulting Group, AB.

Alesina, Alberto, Arnaud Devleeschauwer, William Easterly, Sergio Kurlat, and Romain Wacziarg. 2003. "Fractionalization." *Journal of Economic Growth* 8 (2): 155–94.

Alkire, Sabina, and Maria Emma Santos. 2010. "Acute Multidimensional Poverty: A New Index for Developing Countries." OPHI Working Paper 38, Oxford Policy and Human Development Initiative, Oxford University, Oxford, U.K.

Bueno de Mesquita, Max. 2013. "The Impact of Disarmament, Demobilization and Reintegration on Post-Conflict Elections: The Case of Mozambique and Angola." Master's thesis, Leiden University, The Hague.

Collier, Paul, and Anke Hoeffler. 1998. "On Economic Causes of Civil War." *Oxford Economic Papers* 50 (4): 563–73.

———. 2004. "Aid, Policy and Growth in Post-Conflict Societies." *European Economic Review* 48 (5): 1125–45.

Conley, Timothy G., Christian B. Hansen, and Peter E. Rossi. 2012. "Plausibly Exogenous." *Review of Economics and Statistics* 94 (1): 260–72.

Duranton, Gilles, Peter M. Morrow, and Matthew A. Turner. 2014. "Roads and Trade: Evidence from the US." *Review of Economic Studies* 81 (2): 681–724.

Esteban, Joan, Laura Mayoral, and Debraj Ray. 2012. "Ethnicity and Conflict: An Empirical Study." *American Economic Review* 102 (4): 1310–42.

Esteban, Joan, and Gerald Schneider. 2008. "Polarization and Conflict: Theoretical and Empirical Issues." *Journal of Peace Research* 45 (2): 131–41.

FAO and IIASA (Food and Agriculture Organization of the United Nations and the International Institute for Applied Systems Analysis). 2000. "Global Agro-Ecological Zones (GAEZ)." FAO, Rome; and IIASA, Laxenberg, Austria.

Fearon, James D. 2007. "Economic Development Insurgency and Civil War." In *Institutions and Economic Performance*, edited by Elhanan Helpmann. Cambridge, MA: Harvard University Press.

Felix, Marc L., and Charles Meur. 2001. *Peoples of Africa Atlas: An Enthnolinguistic Atlas.* Brussels: Congo Basin Art History Research Center.

GAO (U.S. Government Accountability Office). 2008. "Afghanistan Reconstruction: Progress Made in Constructing Roads but Assessments for Determining Impact and a Sustainable Maintenance Program Are Needed." GAO-08-689, United States Government Accountability Office, Washington, DC.

Garcia-López, Miquel-Àngel, Adelheid Holl, and Elisabet Viladecans-Marsal. 2013. "Suburbanization and Highways: When the Romans, the Bourbons and the First Cars Still Shape Spanish Cities." IEB Working Paper No. 2013/005, Barcelona.

Ghosh, T., R. L. Powell, C. D. Elvidge, K. E. Baugh, P. C. Sutton, and S. Anderson. 2010. "Shedding Light on the Global Distribution of Economic Activity." *Open Geography Journal* 3 (1): 148–61.

Gurr, Ted Robert. 2000. *Peoples versus States: Minorities at Risk in the New Century.* Washington, DC: United States Institute of Peace Press.

Horowitz, Donald L. 1985. *Ethnic Groups in Conflict.* Berkeley: University of California Press.

Laitin, David D., and Jacob N. Shapiro. 2008. "The Political, Economic, and Organizational Sources of Terrorism." In *Terrorism, Economic Development, and Political Openness*, edited by Philip Keefer and Norman Loayza, 209–230. Cambridge, U.K.: Cambridge University Press.

LandScan. 2006. Global Population Database (2006 release). Oak Ridge National Laboratory, Oak Ridge, Tennessee. http://www.ornl.gov/landscan/.

Martincus, Christian Volpe, Jerónimo Carballo, and Ana Cusolito. 2012. "Routes, Exports, and Employment in Developing Countries: Following the Trace of the Inca Roads." Unpublished, Inter-American Development Bank.

Mauro, Paolo. 1995. "Corruption and Growth." *Quarterly Journal of Economics* 110 (3): 681–712.

Miguel, Edward, Shanker Satyanath, and Ernest Sergenti. 2004. "Economic Shocks and Civil Conflict: An Instrumental Variables Approach." *Journal of Political Economy* 112 (4): 725–53.

Montalvo, Jose G., and Marta Reynal-Querol. 2005. "Ethnic Polarization, Potential Conflict, and Civil Wars." *American Economic Review* 95 (3): 796–816.

MPSMRM, MSP, and ICF International (Ministère du Plan et Suivi de la Mise en œuvre de la Révolution de la Modernité, Ministère de la Santé Publique, and ICF International). 2007. *Democratic Republic of Congo Demographic and Health Survey 2007: Key Findings.* Rockville, MD: MPSMRM, MSP, and ICF International.

Pushak, Nataliya, and Cecilia Briceño-Garmendia. 2011. "The Republic of Congo's Infrastructure: A Continental Perspective." Policy Research Working Paper 5838, World Bank, Washington, DC.

Raleigh, Clionadh, Andrew Linke, Håvard Hegre, and Joakim Karlsen. 2010. "Introducing ACLED: An Armed Conflict Location and Event Dataset Special Data Feature." *Journal of Peace Research* 47 (5): 651–60.

Reynal-Querol, Marta. 2002. "Ethnicity, Political Systems, and Civil Wars." *Journal of Conflict Resolution* 46 (1): 29–54.

Schneider, Gerald, and Nina Wiesehomeier. 2008. "Rules That Matter: Political Institutions and the Diversity-Conflict Nexus." *Journal of Peace Research* 45 (2): 183–203.

UNEP (United Nations Environment Programme). 2011. "UNEP Study Confirms DR Congo's Potential as Environmental Powerhouse but Warns of Critical Threats." UNEP, Nairobi. http://www.unep.org/newscentre/Default.aspx?DocumentID=2656& ArticleID=8890.

Wegenast, Tim C., and Matthias Basedau. 2014. "Ethnic Fractionalization, Natural Resources and Armed Conflict." *Conflict Management and Peace Science* 31 (4): 432–57.

World Bank. 2010. "The Democratic Republic of Congo's Infrastructure: A Continental Perspective." Africa Infrastructure Country Diagnostic, World Bank, Washington, DC.

Chapter **5**

Road Improvements and Deforestation in the Congo Basin Countries

Introduction

Roads bring benefits but may also generate significant environmental impacts, especially in forested areas. Earlier chapters of this report explore different aspects of identifying and measuring the benefits from roads. This chapter turns to the effects of road access on deforestation and biodiversity as estimated in eight Congo Basin countries: Burundi, Cameroon, the Central African Republic, the Democratic Republic of Congo (DRC), Equatorial Guinea, Gabon, the Republic of Congo, and Rwanda. These estimates are then used to simulate the impacts of improvement in a road segment in the Kasese area of Maniema Province, DRC. A focus on the DRC seems warranted in this context because it is the largest country in the Congo Basin and appears most vulnerable to rapid changes in forest cover.

The DRC, the largest country in Sub-Saharan Africa, hosts the second-largest rain forest in the world. The forests of the Congo Basin are home to about 30 million people from more than 100 ethnic groups and remain a crucial livelihood asset, often generating more income for the poor than that obtained from farming (Angelsen et al. 2014). The carbon sequestered by the forests (30–40 gigatons, equal to 8 percent of the world's carbon) gives the DRC the potential to leverage considerable financial resources that are being made available through Reducing Emissions from Deforestation and Forest Degradation (REDD+) and other global initiatives. When fully implemented, REDD+ has the potential to provide benefits such as climate regulation, livelihood support, and biodiversity protection.

But the DRC, as noted in chapter 4, has an immense infrastructure deficit, especially in transport. Only four provincial capitals out of ten can be reached by road from the national capital, Kinshasa, and even by the standards of other low-income countries transport infrastructure remains woefully inadequate (see chapter 4, table 4.1). It is therefore no surprise that road construction and rehabilitation remains a high priority of both the government of the DRC and its major development partners. However, although roads may bring many benefits and are vital for commercializing agriculture, they are often also the precursors to deforestation.

Contribution

A common response to the threat of deforestation stemming from road improvement is the creation of protected areas that severely restrict intrusive structures within demarcated areas. Such strategies often fail to protect critical natural assets because remote protected regions may not coincide with the areas of highest ecological value and also because agricultural and mineral interests invariably overwhelm the limited resources of conservation interests. The political economy of environmental policy suggests two reasons why this might be so. First, the benefits of land conversion are often concentrated, while the environmental costs are diffused, so that collective-action problems render lobbying by environmental groups more difficult. Second, the benefits from land conversion are monetary, while the environmental impacts and costs are typically nonpecuniary, uncertain, and, unlike the benefits, emerge over time, creating further asymmetry in bargaining ability. This chapter develops an approach that could provide additional information that could partly mitigate some of these conflicts and develops a novel metric to identify the impacts of development on forests and areas that are of high ecological value.

Empirical research has provided many useful insights into the determinants of forest clearing. Although econometric work on long-term deforestation drivers is well advanced, previous data problems have limited treatment of economic dynamics to theoretical work and simulations. However, the advent of monthly and annual remote sensing databases has led to the first spatial estimation exercises that explicitly incorporate economic dynamics (Wheeler et al. 2013; Dasgupta et al. 2014). Direct impact studies in Latin America using the new databases have included high-resolution work on new road construction and deforestation in Brazil (Laurance, Goosem, and Laurance 2009) and Bolivia, Panama, Paraguay, and Peru (Reymondin et al. 2013). To the best of the authors' knowledge, the present study is the first extension of such work to Sub-Saharan Africa.

Data and Estimation Framework

This study combines several sources of data to analyze the drivers of deforestation in the Congo Basin. It uses different estimation techniques for robustness checks to estimate the effect of road improvement on deforestation and biodiversity in the Congo Basin.

Road Data

The study uses information for seven Congo Basin countries from digital road maps provided by the Africa Infrastructure Country Diagnostic (AICD). All seven countries are amply represented, with the exception of Gabon for which there is limited information on road conditions. These data are augmented by information on roads for the DRC provided by DeLorme, a mapping and GPS company, and for Equatorial Guinea by DIVA-GIS, a free and open source geographic information system. As table 5.1 shows, the AICD database includes information on the quality of 1,710 segments of 380 roads, with data on both road surface and road conditions for 941 segments.[1]

Forest Clearing Data

This study uses comprehensive and up-to-date data on deforestation rates obtained from Hansen et al. (2013); see box 5.1 for details. Map 5.1 displays gridded estimates of deforestation for the border region of Cameroon, Gabon, the Republic of Congo, and the Central African Republic. Each grid cell is color coded for cumulative percent deforested from 2000 to 2012. Map 5.1 reveals a striking pattern of deforestation along some of the roads in the AICD database. In other cases, however, deforestation is much less pronounced. This study's aim is to better understand the drivers of forest loss and thereby better inform policy decisions.

The estimation exercise in this chapter draws on the insights from previous research to incorporate seven critical determinants of forest clearing in road corridors: road quality, distance from the road, travel time to the nearest market center, the agricultural opportunity value of the land, terrain elevation, legal protection status, and the incidence of violent conflict. Simultaneity bias may be significant in this context, since forest clearing and road placement are jointly

Table 5.1 Congo Basin Roads Data

Country	Roads	Road segments	Segments for which there are data on road surface and condition
Burundi	39	158	37
Cameroon	103	725	397
Central African Republic	30	103	103
Democratic Republic of Congo	121	347	180
Gabon	51	122	7
Republic of Congo	25	155	144
Rwanda	11	100	73
Total	380	1,710	941

Source: Foster and Briceño-Garmendia 2010.

Hansen Pixels of Deforestation

This study uses Hansen et al.'s (2013) published high-resolution data of consistently derived estimates of global forest clearing for the period 2000–12, resampling it to obtain deforestation rates in 2.7 kilometer square grid cells. These data, derived from satellite imagery, are the most comprehensive and current data available on deforestation trends. It represents a considerable advance compared with previous measures of deforestation that were typically based on less reliable and often ad hoc approaches. A disadvantage of the Hansen data is that it cannot distinguish between types of forests, for example, native versus plantation, nor can it provide estimates of forest degradation (that is, thinning of forest cover), often a precursor of deforestation. However, in the Democratic Republic of Congo no significant investment in forest plantation has been made, so the data provide a more accurate description of net forest loss. Each cell contains 8,100 Hansen pixels (30 square meters), so the total counts of Hansen cleared pixels within cells are equivalent to deforestation rates.

Map 5.1 Forest Clearing and Road Networks, 2000–12

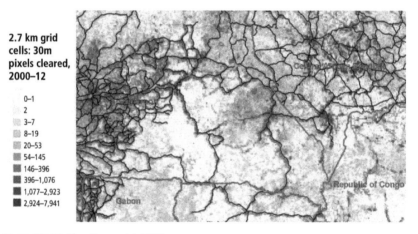

Source: Calculated from Hansen et al. (2013).

determined in a properly specified spatial economic model. The measure of travel time to the nearest urban center along existing roads is treated as endogenous in the estimation, and to correct for potential estimation bias, this study uses Euclidean distance to the nearest urban center as an instrumental variable for travel time to the nearest market (see box 5.2).

BOX 5.2

Estimation Strategy

To determine the impact of road improvements on deforestation, the follow-ing model is estimated:

$$\ln(h_{ijt}) = \alpha_0 + \alpha_1 \ln(d_{ij}) + \alpha_2 \ln(p_i d_{ij}) + \alpha_3 \ln(q_j) + \alpha_4 \ln(m_j)$$
$$+ \alpha_5 \ln(c_i) + \alpha_6 \ln(e_i) + \alpha_7 \ln(v_i) + \sum_{t=1}^{T} \beta_t y_t + \varepsilon_{it}$$

in which
h_{ijt} = deforestation rate of grid cell i, along road segment j, at year t
d_{ij} = distance of the centroid of cell i from road segment j
p_i = legal protection status of cell i (dummy = 1 if protected, 0 otherwise)
q_j = quality of road segment j
m_j = travel time from the midpoint of segment j to the nearest urban center
    (that is, the market)
c_i = average agricultural opportunity cost of cell i
e_i = average elevation of cell i
v_i = violent conflict incidence in cell i
y_t = year fixed effects (dummy = 1 if year t, 0 otherwise)
    (See annex 5A for details on these variables.)

Forest clearing and road placement are jointly determined; therefore, travel time to the nearest market along existing roads (m_j) should be treated as endogenous. To account for this, the Euclidean distance from segment j to the nearest market is used to instrument for travel time.[a] Population density is excluded from these regressions because it is likely endogenous and cor-related with urban agglomerations (markets) as well as land opportunity, both of which are included as explanatory variables.

The model is estimated by two-stage least squares (2SLS), generalized least squares with instrumental variables (GLS-IV), and robust regression IV. The GLS-IV estimates (which adjust the covariance matrix for road-specific error variances) are the preferred estimates.

a. Note that the travel time to the nearest market is a different measure from that used in the earlier chapters for "transport cost to market." The Euclidean distance, instead of the natural path, is used as an IV because of the lack of data needed to compute the natural path instruments for all of the Congo Basin countries.

Estimation Results

Table 5.2 presents estimates of the impact of road access on forest clearing in the Congo Basin using several estimation techniques. The first two columns present ordinary least squares (OLS) results for travel time and Euclidean distance. Columns 3 through 5 present two-stage least squares (2SLS), generalized least squares with instrumental variables (GLS-IV), and robust regression instrumental variable (IV) results, respectively. The GLS-IV estimates are the

Table 5.2 Regression Results
(all nondummy variables in log form)

	(1)	(2)	(3)	(4)	(5)
Dependent variable: Cumulative Hansen pixels cleared	OLS	OLS-alternate market distance	2SLS	GLS-IV	Robust IV
Distance from road	−0.535***	−0.672***	−0.672***	−0.672***	−0.658***
	(68.68)	(−102.73)	(−102.73)	(−24.18)	(−102.11)
Protected area × Distance from road	−0.132***	−0.164***	−0.164***	−0.164***	−0.162***
	(22.59)	(−28.32)	(−28.32)	(−4.28)	(−28.48)
Road condition	0.209***	0.205***	0.205***	0.205**	0.219***
	(14.80)	(14.48)	(14.48)	(2.15)	(15.71)
Travel time to nearest urban center	−0.471***		−0.628***	−0.628***	−0.583***
	(−37.35)		(−30.57)	(−4.84)	(−28.81)
Euclidian distance to market		−0.236***			
		(−30.57)			
Land opportunity value	0.031***	0.069***	0.069***	0.069**	0.067***
	(5.95)	(14.01)	(14.01)	(2.04)	(13.71)
Elevation	−0.185***	−0.258***	−0.258***	−0.258***	−0.244***
	(−15.79)	(−22.41)	(−22.41)	(−3.23)	(−21.55)
Conflict intensity (1997–2007)	0.025***	0.031***	0.031***	0.031	0.018***
	(8.52)	(10.65)	(10.65)	(1.64)	(6.45)
Constant	5.339***	4.252***	6.985***	6.985***	6.668***
	(46.48)	(39.65)	(45.28)	(7.06)	(43.88)
Country fixed effects	Yes	Yes	Yes	Yes	Yes
Year fixed effects	Yes	Yes	Yes	Yes	Yes
Observations	44,743	44,743	44,743	44,743	44,743
R^2	0.56	0.56	0.56	0.56	0.56

Sources: Hansen et al. 2013; and calculations.
Note: 2SLS = two-stage least squares; GLS = generalized least squares; IV = instrumental variable; OLS = ordinary least squares. *t*-statistics in parentheses.
*** $p < 0.01$, ** $p < 0.05$, * $p < 0.10$.

preferred ones, and the following discussion focuses on these. In brief, these estimates are remarkably robust and stable across different statistical specifications in the regressions. The results suggest that forest clearing intensity declines on average with (1) distance from roads and markets, (2) proximity to protected areas, and (3) less accessible terrain (for example, higher elevation); and increases on average with (4) improvements in road conditions, (5) the agricultural value of land (opportunity cost), and (6) conflict intensity.[2]

The main variable of interest in this analysis is the distance from the road. The coefficient estimate is consistent and highly significant across the estimators. All else equal, a 10 percent increase in distance from the road leads to a 6.7 percent decline in deforestation intensity in unprotected areas (column 4 of table 5.2). If the area is protected, then a 10 percent increase in distance from the road would lead to an 8.4 percent decrease in deforestation. Another coefficient that warrants brief explanation is conflict. In columns 3 and 5, conflict is found to have a significant and positive influence on deforestation rates. However, conflict is not significant in the GLS-IV estimates, so the finding is less robust. Nonetheless, these results are informative in that they represent—to the best of the authors' knowledge—the first statistical estimate of the impact of violent conflict on deforestation. It has often been conjectured (but not empirically tested) that conflict, by slowing development, has provided tacit protection to the Congo Basin forests (Megevand 2013). Conversely, displacement and lack of opportunity created by conflict could contribute to higher levels of deforestation by increasing the levels of forest dependence. Endogeneity may be a further concern if the level of forest cover influences the outcomes of conflict. These findings suggest the need for further deeper research on the effects of conflict on deforestation.

The yearly dummy variables (not shown for the sake of brevity) increase steadily but at a decreasing rate during the period 2000–12. All country dummies (also not shown in table 5.2) except those for the Central African Republic are highly significant (and highly varied) in the 2SLS and robust IV results, although the GLS-IV results suggest that the large positive estimate for the DRC is the most robust in the set.[3] (For detailed discussion of these results, see annex 5B.)

Two patterns are noteworthy. First, upgrading roads from very poor to good condition produces near-complete deforestation within a narrow corridor (of about 1.0–1.5 kilometer radius) straddling the road. Second, the impact is nonlinear and deforestation intensity falls very rapidly as distance from the road increases. These patterns are further discussed in the simulations below.

Simulated Implications for Forest Clearing

This section explores the implications of the GLS-IV estimates in table 5.2 on deforestation rates. It illustrates the micro-implications using a road segment in the Kasese area of Maniema Province, DRC. A simulation of a road segment upgrade is performed, which affects the underlying data in two important ways. Starting with the current road condition (in which much of the simulated road is rated as 1, or very poor), the condition for the entire road segment is incrementally improved to 2 (poor), 3 (fair), and 4 (good). This improvement in the road condition reduces the travel time to the nearest urban center for routes that pass through these road segments. Thus, the coefficients for two variables— "travel time to nearest urban center" and "road condition" are used to generate an estimate for the amount of additional deforestation that might occur if this road segment were improved.[4]

Figure 5.1 presents forest clearing results for the Kasese road segment as its condition improves from very poor (1) to good (4). With the road in very poor condition, 34 percent of previously forested land is cleared within 200 meters of the road. Farther away, clearing declines to 11 percent at 1 kilometer, 7 percent at 2 kilometers, and 2 percent at 10 kilometers. When the road segment's condition is improved to fair, the percentage of clearing predicted at 200 meters, 1 kilometer, 2 kilometers, and 10 kilometers is 66 percent, 22 percent, 14 percent, and 5 percent, respectively. With further upgrading to good condition, clearing

Figure 5.1 Effect of Road Quality on Forest Clearing Intensity

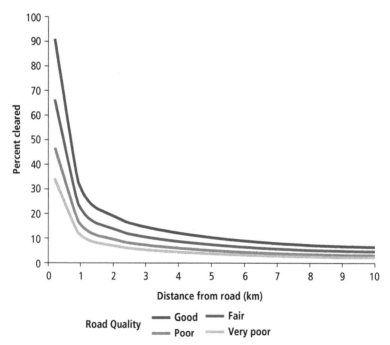

Source: World Bank calculations.

at the four distances increases to 91 percent, 30 percent, 19 percent, and 6 percent. In sum, it can be observed that upgrading from very poor to good produces near-complete deforestation within a narrow corridor straddling the road.

Deforestation intensity falls rapidly as distance from the road increases, as illustrated in figure 5.1. Despite this decline, the overall spatial magnitude of the impact of road improvement is striking. To illustrate, suppose that 5 percent clearing was adopted as the criterion for identifying pixels at the outer margin of corridor deforestation. For the road in poor condition, this yields an affected corridor 6 kilometers wide (3 kilometers on either side of the road). Upgrading to good condition widens the affected corridor to 26 kilometers. Map 5.2 presents the results for the DRC, color coded by predicted change in the deforestation rate along road corridors 2 kilometers wide.

Map 5.3 displays changes in mean deforestation rates produced by road upgrading to level 4 (good). The results are presented at the subdistrict level. For each subdistrict, mean deforestation rates are computed before and after upgrading, for pixels whose centroids lie within 2-kilometer corridors straddling all roads in the subdistrict. Attention is restricted to 2 kilometers because this is the range over which the greatest impact is predicted to occur. The heaviest impacts (increases of 13.7– 22.4 percent) are evident in west Orientale, east Equateur, central Kasaï-Occidental, northeast Kasaï-Oriental, and

Map 5.2 Eastern DRC: Change in Percent Clearing along Roads with Upgrading

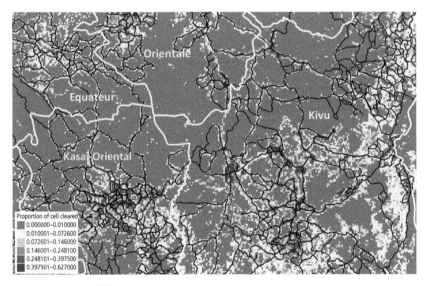

Sources: Hansen et al. 2013; and calculations.
Note: DRC = Democratic Republic of Congo.

Map 5.3 Changes in Road Corridor Deforestation in the DRC with Generalized Upgrading

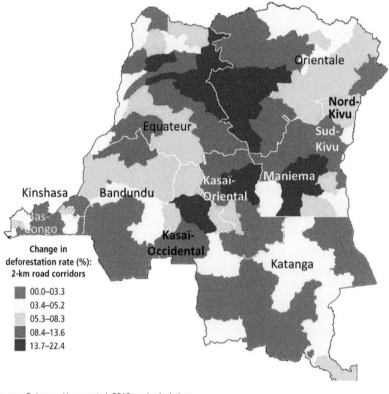

Sources: DeLorme; Hansen et al. 2013; and calculations.
Note: DRC = Democratic Republic of Congo.

central Maniema. Adjacent areas in all five provinces also have significant impacts (8.4 percent to 13.6 percent). The larger impacts are concentrated in relatively isolated rain forest areas with poor roads because market access for these areas would be the most improved by upgrading. Overall, the results indicate that 10–20 percent increases in deforestation would be common after upgrading in rain forest road corridors.

Deforestation Effects on Biodiversity

Not all forested land is of uniform ecological value, nor is it of uniform economic value. Deforestation may reduce or eliminate critical habitats of endangered animal and plant species, so identifying these areas is useful for decision makers.

Accordingly, this chapter develops a novel metric to identify areas that are of high ecological value and where species are at high risk of extinction. Results from efforts to develop a single index of ecological value remain elusive and are beset with difficulties of enumeration and measurement. Nor has a consensus emerged on the weighting of different risks and species. To address these issues, this chapter develops a "gradient approach" with a composite index that incorporates information from a variety of sources into a single index. This approach has practical policy merit. Some road corridors will be built in areas of modest ecological concern, whereas others will pass through areas of higher value. An ecological gradient strategy, henceforth "ecogradient," can be used to minimize ecological damage by favoring road improvements in areas of modest concentration of valuable biodiversity.

A useful ecogradient database should have tractable size while incorporating sufficient spatial resolution for realistic local applications. This exercise employs a 5-kilometer spatial grid that covers the Congo Basin with 133,782 square cells. A composite gradient is developed that includes measures of endemicity, three measures of extinction risk from the International Union for Conservation of Nature (IUCN), and a combined measure of phylogenetic significance and extinction probabilities (from Isaac et al. 2007). Each index generates a different set of priorities. A cautious approach is used—the indicator that generates the highest threat level is used. To implement this degree of caution, the indices for comparability are normalized[5] and the maximum index (risk) value is selected as the ecogradient measure for the cell. This approach gives parity to alternative vulnerability indicators and always picks the indicator that generates the highest threat level relative to the threat level of other indicators.

An ecogradient measure based on species vulnerability alone provides an incomplete accounting of ecological values and functions. A more comprehensive measure would need to incorporate biomes. Using the World Wildlife Fund (WWF) classification of ecoregions, a vulnerability index is derived that measures the amount of an ecoregion in a given area. Accordingly, this study adopts the ecoregion as a general proxy for distinctive plant, insect, and animal species that are not represented in the range maps provided by the IUCN and BirdLife International.

The method for incorporating WWF ecoregions (shown globally in map 5C.1 and for the Congo Basin in map 5C.2 in annex 5C) resembles this study's treatment of species endemicity. For the group of selected countries, the percentage of total area accounted for by each ecoregion is computed. Each ecoregion's vulnerability index is then computed as the inverse of its area share, and the appropriate index value is assigned to each pixel in the Congo Basin countries. This method assigns high values to pixels in smaller ecoregions, where clearing single pixels may pose more significant threats to biome integrity.

No consensus exists on relative weights for animal species and ecoregion indices in a composite index. As before, a conservative accommodation could have been provided by normalizing both indices to a range of 0–100 for each pixel and choosing the larger index value as the final measure. However, the ecoregion index is far more skewed than the species index. As a consequence, choosing maximum values for each normalized index pair will select the species index in the great majority of cases. For a more robust and balanced measure, ranks—measured as percentiles—are used instead, and the greater percentile value for each pixel is chosen.

Map 5.4 thus combines both sets of information, displaying the distribution of the species-ecoregion index in the Congo Basin countries. One striking

Map 5.4 Composite Species-Ecoregion Index, Congo Basin Countries

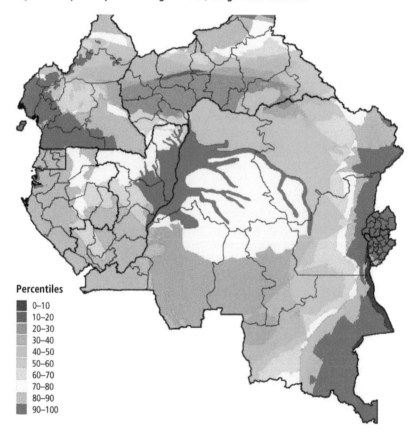

Percentiles

- 0–10
- 10–20
- 20–30
- 30–40
- 40–50
- 50–60
- 60–70
- 70–80
- 80–90
- 90–100

Sources: World Wildlife Fund (https://www.worldwildlife.org/pages/conservation-science-data-and-tools); and calculations.

feature is the blue-green (0–50) band that arcs from northern Cameroon to eastern DRC and back to southern DRC. Another is the prominent clustering of very high values in western Cameroon, along the border between the Republic of Congo and the DRC, and along the eastern margin of the basin.

For the purposes of this chapter, the most important message in the results is the striking non-uniformity of ecological vulnerability across forested areas in the Congo Basin countries. By implication, a full assessment of the benefits and costs of road upgrading should go beyond simple measurement of forest loss to an evaluation of the potential impact of that loss on biological diversity.

Road Improvement Revisited—The Stakes for Vulnerable Areas

In this section, the three strands of research in this report are joined by combining the economic benefits and predicted deforestation from road upgrading with the composite index of ecological vulnerability. This study considers a hypothetical example of the upgrading of a road close to the globally renowned Virunga National Park, home to the critically endangered mountain gorilla, but also located in a heavily populated part of the Congo Basin in an area with high and persistent incidences of conflict. Specifically, this study considers the consequences of upgrading the entire road from Goma to Bunia to paved and good condition. Currently, about 140 kilometers of this road is paved, but in poor condition, while the remaining 385 kilometers (200 kilometers in fair or good condition and the rest in poor condition) is unpaved.

The total annual increase in GDP from the improvement of the Goma to Bunia road is predicted to be US$7.29 million, arguably a modest increase, no doubt reflecting the limited productivity of this conflict-prone region. The spatial distribution of these benefits is shown in map 5.5. Not surprisingly, the highest benefits are clustered around those roads surrounding cities.

Next, the deforestation regression estimates are used to chart the changes in forest cover in the same region. Accordingly, the estimates predict an average increase in deforestation of 2.6 percent for affected cells, ranging from 0 percent to 45 percent depending on distance to road, distance to the nearest market, and initial condition of the nearest road segment, unsurprisingly closely mirroring the changes in transport costs. The spatial distribution of the predicted deforestation is shown in map 5.6. Overlaying the biodiversity gradient, as shown in map 5.7, indicates that this is an area of high biodiversity significance, suggesting the need for caution and improved management to accompany the change in road condition, especially given the vast and as yet unrealized tourism potential and revenues that could be generated from this region once peace dividends begin to flow.

Map 5.5 Changes in GDP from Road Upgrade, Goma to Bunia

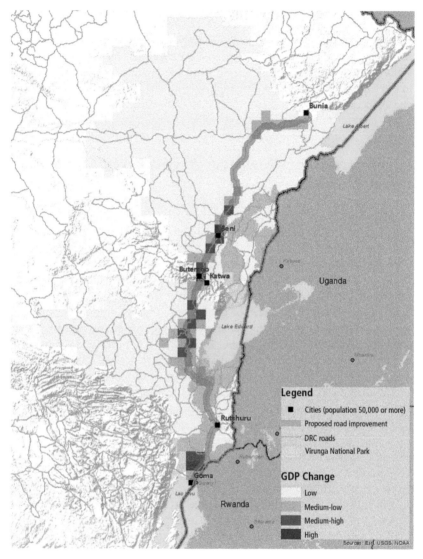

Legend

- ■ Cities (population 50,000 or more)
-   Proposed road improvement
-   DRC roads
-   Virunga National Park

GDP Change

-   Low
-   Medium-low
-   Medium-high
-   High

Source: Esri, USGS, NOAA

Sources: DeLorme; Ghosh et al. 2010; and calculations.
Note: DRC = Democratic Republic of Congo.

Map 5.6 Changes in Forest Cover from Road Upgrade, Goma to Bunia

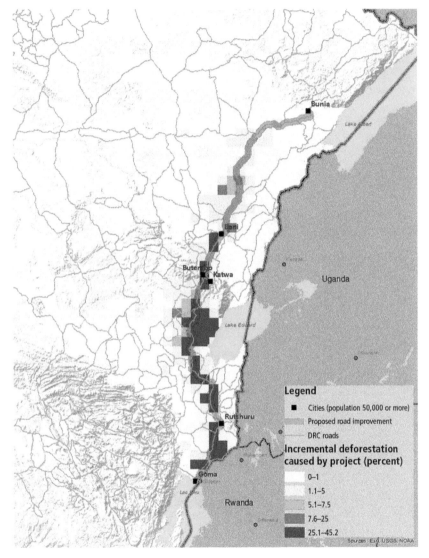

Sources: DeLorme; Hansen et al. 2013; and calculations.
Note: DRC = Democratic Republic of Congo.

Map 5.7 Biodiversity and the Affected Area

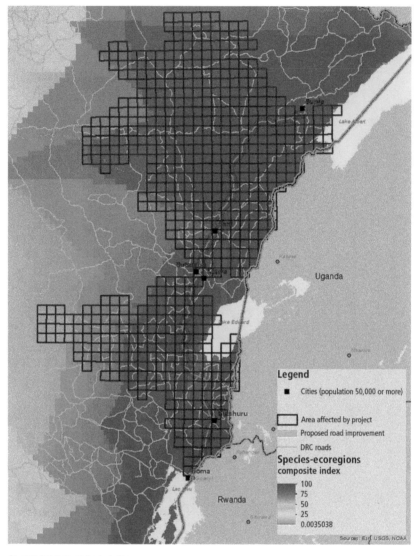

Sources: DeLorme; and calculations.
Note: DRC = Democratic Republic of Congo.

Summary and Conclusion

This chapter uses high-resolution spatial data for eight Congo Basin countries to develop and estimate an econometric model of deforestation that incorporates the economics of road improvement. Overall, the results suggest that plans for road investments need to consider the possible physical impacts on forests and ecosystems from the very beginning. Sequential decision making, whereby location decisions occur first followed by environmental impact assessments, can lead to economically less favorable outcomes that can be avoided by careful upstream planning. Both new data and new techniques are now available that can be used to identify areas of opportunity, risk, and potential for REDD+ financing. This study finds large, highly significant effects of road upgrading on the intensity and extent of forest clearing in road corridors. In addition, the results from the empirical analysis highlight the powerful roles played in proprietors' forest-clearing decisions by market access, land opportunity values, official protection status, and topography. The results provide the first estimate of the impact of violent conflict on deforestation in Sub-Saharan Africa.

The econometric estimates predict the impact of generalized road upgrading on forest clearing along road corridors in the Congo Basin and illustrate the results with a detailed assessment of impacts in the DRC. Predicted effects in road corridors vary widely with previous road conditions and locational economics, but increases of 10–20 percent are typical. In addition, many corridors have significant extensions in the outer margin of forest clearing.

After investigating the impact of road improvements on forest clearing, this study extends the analysis to potential ecological impacts. Using spatially formatted databases from the IUCN, BirdLife International, and the WWF, a pixel-level measure of biodiversity risk is constructed for thousands of animal species and plant biomes. The resulting high-resolution map reveals a complex geographic pattern of ecological vulnerability. This study overlays the basin-wide road network on this vulnerability map to provide a first-order guide to risk assessment for proposed road corridor improvements.

In the concluding stage of the analysis, this study combines the ecological risk indicator with pixel-level predictions of forest clearing produced by road upgrading. The result is a high-resolution map of expected risks from road upgrading in road segments, corridors, and regional networks. Predicted deforestation is not correlated with the vulnerability indicator, so the variation in the two indicators separately is compounded in the combined measure. This study illustrates the implications for the eastern DRC in a detailed assessment.

Overall, the results cast doubt on the utility of broad generalizations about the impact of road upgrading on deforestation, biomes, and vulnerable indigenous communities. The high-resolution spatial assessment finds impacts as varied as the economic, social, and ecological conditions that prevail in different road corridors. By implication, road improvement planning in tropical forest regions is unlikely to maximize welfare unless it anticipates and incorporates such impacts.

Annex 5A Data and Variable Definitions

Data sets for individual years in the period 2000–12 are stacked, allowing for different mean clearing in grid cells by year. All variables are calculated as centroid values for 2.7 kilometer Hansen grid cells. Data and sources are as follows:

Hansen pixels cleared: 2.7 arcsecond grid cell summing up the number of 30 meter pixels identified by Hansen et al. (2013) as cleared, by year, for the period 2000–12.

Distance from road segment: Distance from the centroid of each grid cell to the nearest road, calculated in ArcGIS 10.

Legal protection status: 1 if the parcel is in a protected area identified by the World Database on Protected Areas (WDPA); 0 otherwise. The WDPA shapefile was downloaded from http://www.protectedplanet.net/.

Condition of the road segment: Road surface type and road condition, obtained from the AICD. Surface types are earth, gravel, and asphalt. Road conditions are identified as good, fair, poor, and very poor. Each variable is translated into ordered cardinal values for estimation: Surface: dirt (1), gravel (2), asphalt (3); Condition: very poor (1), poor (2), fair (3), good (4).

Travel time: Time from a road segment's mean point to the nearest urban center with a population of 50,000 or greater, as estimated by Uchida and Nelson (2009). Raster resolution: 0.0083 decimal degrees.

Agricultural opportunity value: Mean value for a grid cell, calculated from the high-resolution global grid developed by Deveny et al. (2009). Raster resolution: 0.0025 decimal degrees.

Elevation: Average elevation for a grid cell, calculated from the Consultative Group for International Agricultural Research–Shuttle Radar Topography Mission data set (3 second resolution), aggregated to 30 second resolution by DIVA-GIS (http://www.diva-gis.org/gdata).

Conflict incidence: Armed conflict fatalities per unit area, 1997–2007, calculated in chapter 3 at 0.017 decimal degrees resolution from data in the Armed Conflict Location Events Dataset (Raleigh et al. 2010).

Annex 5B Empirical Results

The first two columns in table 5.2 present OLS results for travel time and Euclidean distance. Columns 3 through 5 present three results that employ Euclidean distance as an instrument for travel time: standard 2SLS, GLS-IV with covariance matrix adjustments for road-specific error variances, and robust IV regression.

The presented results are distilled from numerous experiments that tested the interactions of distance to road with road quality, travel time, agricultural opportunity value, elevation, and protection status. The experiments were critical for determining whether these variables affect the slope of the relationship between forest clearing and distance from the road. Only protection status has a strong, consistent effect on the slope, so the interaction of protection and distance is retained in table 5.2. Country dummies are also included to control for systematic differences in forest-clearing intensity. Previous experimentation also revealed that road surface type has no significance for forest clearing, controlling for road condition; therefore, this variable is excluded from the final regressions.

All variables have generally high significance, and their signs are consistent with expectations. Among the IV regressions, the GLS corrections for non-uniform error variances across roads result in substantially higher standard errors (and lower t-statistics), but, with the exception of conflict intensity, all variables retain significance levels of 95 percent or higher. The GLS-IV estimates are used for the discussion of results.

The most critical variable for this exercise is distance from the road. In table 5.3, the results for this variable are strong and consistent across estimators: all else equal (in GLS-IV), forest clearing intensity declines 0.67 percent with each 1 percent increase in distance. The result for protection status indicates that the relationship between clearing intensity and distance steepens significantly in protected areas.

The results for road quality are also consistent across specifications, and for numerous experiments that test interactions with distance from the road. Road surface (earth, gravel, asphalt) never has a significant effect,[6] but the impact of road condition (very poor, poor, fair, good) is large and highly significant. Neither road surface nor road condition interacts significantly with distance from the road.

Among other regression variables, forest clearing is negatively related to travel time to the nearest market center (elasticity −0.63 in GLS-IV) and elevation (elasticity −0.26), and positively related to land opportunity value (0.07) and conflict intensity (0.03). As expected, IV results for travel time to the nearest urban center differ markedly from the OLS results.

Annex 5C Biodiversity Measurement

Species density provides critical information for developing ecogradients, but at least two other elements are needed:

- Geographic vulnerability, which can be proxied by endemicity, that is, the proportion of each species' range that lies within each grid cell. Species that reside in very few grid cells may be particularly vulnerable to habitat encroachment. Endemicity can be computed from the database constructed for this chapter.

- A measure of extinction risk that adds insights from the international scientific community. Recent work by Mooers, Faith, and Maddison (2008) explicitly models the relationship between extinction probability and the risk indicator that is provided for each species in the IUCN and BirdLife International databases.[7]

Endemicity

The study's ecogradient database includes endemicity, that is, the percentage of each species' range that is found in each grid cell. Endemicity treats all species equally at the global level, since each species has a total count of 1. Total endemicity for each grid cell—the sum of its species' endemicity measures—assigns higher values to cells inhabited by species whose ranges are relatively limited. By implication, forest clearing in higher-value cells may be particularly destructive for remaining critical habitat.

Extinction Risk

Species differ in vulnerability for many reasons that are not captured by the endemicity measure. To incorporate these factors, the threat status code assigned to each species by the IUCN's Red List is used. An appropriate measure of vulnerability in this context is extinction risk, so Red List status codes are converted to extinction probabilities using the methodology of Mooers, Faith, and Maddison (2008). For species indicator construction, these probabilities are normalized so that a weight of 1 is assigned to species in the highest category (Critically Endangered). Table 5C.1 tabulates conversions from Red List codes to normalized species weights, using four probability assignments. The first column draws on recent work by Isaac et al. (2007), who combine a direct extinction risk measure with a measure of each species' isolation on a phylogenetic tree.[8] Columns 2 through 4 display three IUCN estimates that are employed to derive measures of extinction probability during the next 50, 100, and 500 years.

Table 5C.1 Normalized Species Aggregation Weights

		Normalized extinction probabilities			
		IUCN: Future years[a]			
IUCN Code	Status	(1) Isaac[b]	(2) 50 years	(3) 100 years	(4) 500 years
CR	Critically endangered	1.00000	1.00000	1.00000	1.00000
EN	Endangered	0.50000	0.43299	0.66770	0.99600
VU	Vulnerable	0.25000	0.05155	0.10010	0.39000
NT	Near threatened	0.12500	0.00412	0.01000	0.02000
LC	Least concern	0.06250	0.00005	0.00010	0.00050
	Rounded weight ratios				
	CR:EN	2	2	1	1
	CR:VU	4	19	10	3
	CR:NT	8	243	100	50
	CR:LC	16	20,000	10,000	2,000

Sources: a. Mooers, Faith, and Maddison 2008. b. Calculations by Mooers, Faith, and Maddison (2008), based on Isaac et al. (2007).
Note: IUCN = International Union for Conservation of Nature.

Table 5C.2 Implications of Alternative Weighting Schemes

			Total scores			
			(1)	IUCN extinction probabilities: Future years		
Area	Species count	Status (uniform within areas)	Isaac et al. 2007	(2) 50 years	(3) 100 years	(4) 500 years
A	2	Critically endangered	2	2	2	2
B	20,000	Least concern	1,250	1	2	10

Source: World Bank calculations.
Note: IUCN = International Union for Conservation of Nature.

Table 5C.1 shows that Isaac et al.'s (2007) inclusion of the phylogenetic isolation factor changes the weight ratios substantially, particularly for species in the lowest threat category (Least Concern). The implications for hypothetical areas A and B are explored in table 5C.2. Area A is populated by only two species, both rated as Critically Endangered. Area B is populated by 20,000 species, but all are rated as of Least Concern. This study's extinction risk indicator for each area is the sum of normalized extinction probabilities for resident species. Assignment of weights for Mooers, Faith, and Maddison's (2008) IUCN-derived 50-year extinction probabilities yields a total risk indicator of 2 for A—twice the total for B, because each Critically Endangered species is weight-equivalent to 20,000 Least Concern species. In contrast, assignment of the Isaac weights yields an overall risk rating for B (1,250) that is 625 times greater than the rating for A (2), because each Critically Endangered species is weight-equivalent to 16 species of Least Concern. The other two cases are intermediate, but far closer to the 50-year IUCN case.

Composite Ecogradients

With such potentially huge differences in metrics, an ecogradient methodology must accommodate different risk-weighting schemes in a consistent and plausible way. In addition, it seems highly unlikely that the weights assigned to species classes (amphibians, birds, mammals, reptiles) by the conservation community would be proportional to their grossly unequal representation in overall species counts. Finally, conservation specialists are not likely to agree on the relative importance that should be assigned to endemicity and extinction risk in composite ecogradients.

To accommodate these diverse concerns, a conservative strategy for ecogradient construction is adopted. First, all measures of endemicity and extinction risk are divided by their maximum values and multiplied by 100 to create indexes in the range 0–100. This step ensures comparability in measurement. Then, for each grid cell, the maximum index value is selected as the ecogradient measure for the cell. This approach ensures parity treatment for all animal species and alternative vulnerability indicators in determining composite ecogradients.

Incorporating Biomes

An ecogradient measure based on animal species alone provides an incomplete accounting of biodiversity. A more complete measure would incorporate plants and insects, using indices similar to those that this study has developed for extinction. Although no such indices exist at the requisite geographic scale, the WWF has provided a first approximation by segmenting the world into 825 terrestrial ecoregions (map 5C.1). The WWF defines an ecoregion as "a large unit of land or water containing a geographically distinct assemblage of species, natural communities, and environmental conditions."[9] Accordingly, this study adopts the ecoregion as a general proxy for distinctive plant and insect species, as well as animal species that are not represented in the range maps provided by the IUCN and BirdLife International.

The method for incorporating WWF ecoregions (shown globally in map 5C.1, and for the Congo Basin in map 5C.2) resembles this study's treatment of species endemicity. For this exercise, all ecoregions in the Congo Basin countries are identified. Geographic coverage is then extended to surrounding countries that include parts of those ecoregions. Map 5C.3 identifies these geographies, which extend from Senegal to South Sudan in the north, and from Angola to Malawi in the south. For the group of selected countries, the percentage of total area accounted for by each ecoregion is computed.

Map 5C.1 World Wildlife Fund Terrestrial Ecoregions

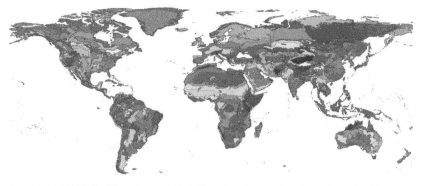

Source: World Wildlife Fund (https://www.worldwildlife.org/pages/conservation-science-data-and-tools).
Note: Colors represent 825 terrestrial ecoregions.

Map 5C.2 World Wildlife Fund Terrestrial Ecoregions in the Congo Basin Countries

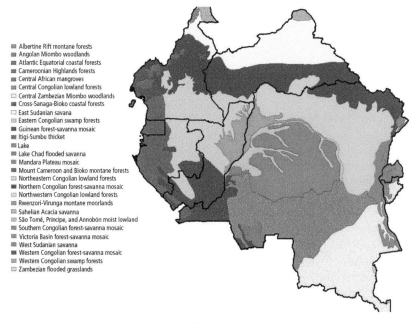

Albertine Rift montane forests
Angolan Miombo woodlands
Atlantic Equatorial coastal forests
Cameroonian Highlands forests
Central African mangroves
Central Congolian lowland forests
Central Zambezian Miombo woodlands
Cross-Sanaga-Bioko coastal forests
East Sudanian savana
Eastern Congolian swamp forests
Guinean forest-savanna mosaic
Itigi-Sumbu thicket
Lake
Lake Chad flooded savanna
Mandara Plateau mosaic
Mount Cameroon and Bioko montane forests
Northeastern Congolian lowland forests
Northern Congolian forest-savanna mosaic
Northwestern Congolian lowland forests
Rwenzori-Virunga montane moorlands
Sahelian Acacia savanna
São Tomé, Príncipe, and Annobón moist lowland
Southern Congolian forest-savanna mosaic
Victoria Basin forest-savanna mosaic
West Sudanian savanna
Western Congolian forest-savanna mosaic
Western Congolian swamp forests
Zambezian flooded grasslands

Sources: World Wildlife Fund (https://www.worldwildlife.org/pages/conservation-science-data-and-tools); and
calculations.

Map 5C.3 World Wildlife Fund Ecoregions in the Greater Congo Basin Region

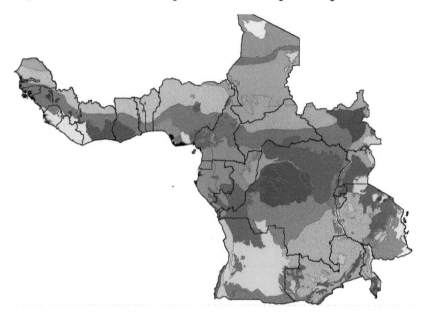

Sources: World Wildlife Fund (https://www.worldwildlife.org/pages/conservation-science-data-and-tools); and calculations.

The area's vulnerability index is then computed as the inverse of its area share, and the appropriate index value is assigned to each pixel in the Congo Basin countries. This method assigns high values to pixels in smaller ecoregions, where clearing single pixels may pose more significant threats to biome integrity.

A Composite Biodiversity Indicator

No consensus exists on relative weights for animal species and ecoregion indices in a composite index. As before, a conservative accommodation could have been provided by normalizing both indices to range 0–100 for each pixel and choosing the larger index value as the final measure. However, the ecoregion index is far more skewed than the species index. As a consequence, choosing maximum values for each normalized index pair will select the species index in the great majority of cases. For a more robust and balanced measure, ranks, measured as percentiles, are used instead, and the greater percentile value for each pixel is chosen.

Map 5C.4 Eastern DRC: Road Networks and Vulnerability

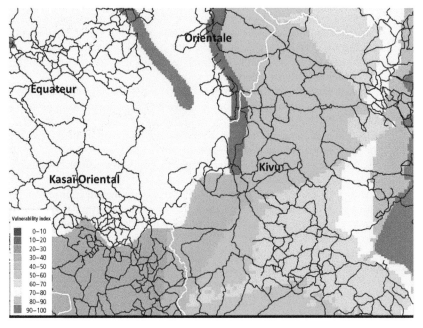

Sources: World Wildlife Fund (https://www.worldwildlife.org/pages/conservation-science-data-and-tool); and calculations.
Note: DRC = Democratic Republic of Congo.

Notes

1. Among road segments in the data that are rated (good, fair, poor, very poor), 32.6 percent are asphalt, 26.7 percent gravel, and 40.7 percent dirt. Road conditions vary widely across segments, with 30.6 percent rated good, 25.6 percent fair, 35.7 percent poor, and 8.1 percent very poor. Higher-quality roads and asphalt roads appear to be well represented, providing useful information for the econometric analysis.
2. Population is not included as an explanatory variable because of endogeneity issues. In fact, population densities are captured in the market-urban variable.
3. Given the robustness of the DRC dummy result, this study also tested the model separately for the DRC. In comparison with table 5.2, identical signs and significance levels are found for distance from the road, its interaction with protected-area status, road condition, travel time to the nearest urban center, and elevation. The DRC parameter estimate for road condition is effectively identical to the Congo Basin estimate in table 5.2.
4. The implications of the GLS-IV results are explored by employing a three-step approach. First, the regression variables are used to predict forest clearing $(\hat{f}_0)$ for

520,732 2.7 kilometer pixels in the eight Congo Basin countries. Where road condition estimates are missing (most often for rural tertiary roads), very poor conditions are assumed (value 1). Second, the effect of upgrading is simulated by progressively resetting all road conditions to higher values (2, 3, 4) and repredicting forest clearing ($\widehat{f}_i$) for all pixels. Third, to ensure consistency with the initial pattern of deforestation, the ratio [$\rho = (\widehat{f}_1)/(\widehat{f}_0)$] is formed and multiplied by actual clearing (f) to obtain the final prediction result [$\widehat{f} = \rho f$].

5. The normalization process is as follows: Let x_i (where $i = 1,2,...,n$) be the set of index numbers and let x_{max} be the highest valued element. The normalization process is $xi/xmax$.

6. The effect of road surface using categorical variables has also been tested for, as well as the previously described cardinal measure.

7. The IUCN's current classification categories are Critically Endangered, Endangered, Vulnerable, Near Threatened, and Least Concern.

8. A phylogenetic tree is a branching tree diagram that traces the evolutionary descent of different species from a common ancestor. Species in sparse (isolated) branches of a phylogenetic tree are relatively unique, given that they share common descent patterns with few other species.

9. Complete information about the WWF terrestrial ecoregions is available online at http://wwf.panda.org/about_our_earth/ecoregions/.

References

Angelsen, Arild, Pamela Jagger, Ronnie Babigumira, Brian Belcher, Nicholas J. Hogarth, Simone Bauch, Jan Börner, Carstten Smith-Hall, and Sven Wunder. 2014. "Environmental Income and Rural Livelihoods: A Global-Comparative Analysis." *World Development* 64 (Suppl 1): S12–S28.

Dasgupta, Susmita, Daniel Hammer, Robin Kraft, and David Wheeler. 2014. "Vyaghranomics in Space and Time: Estimating Habitat Threats for Bengal, Indochinese, Malayan and Sumatran Tigers." *Journal of Policy Modeling* 36 (3): 433–53.

Deveny, A., J. Nackoney, N. Purvis, R. Kopp, E. Myers, M. Macauley, M. Obersteiner, G. Kindermann, M. Gusti, and A. Stevenson. 2009. *Forest Carbon Index: The Geography of Forests in Climate Solutions*. Washington, DC: Resources for the Future and Climate Advisers.

Ghosh, T., R. L. Powell, C. D. Elvidge, K. E. Baugh, P. C. Sutton, and S. Anderson. 2010. "Shedding Light on the Global Distribution of Economic Activity." *Open Geography Journal* 3 (1): 148–61.

Foster, Vivien, and Cecilia Briceño-Garmendia. 2010. "Africa's Infrastructure: A Time for Transformation." Africa Development Forum, World Bank, Washington, DC.

Hansen, M. C., P. V. Potapov, R. Moore, M. Hancher, S. A. Turubanova, A. Tyukavina, D. Thau, S. V. Stehman, S. J. Goetz, T. R. Loveland, A. Kommareddy, A. Egorov, L. Chini, C. O. Justice, and J. R. G. Townshend. 2013. "High-Resolution Global Maps of 21st-Century Forest Cover Change." *Science* 342 (6160): 850–53.

Isaac, N., S. Turvey, B. Collen, C. Waterman, and J. Baillie. 2007. "Mammals on the EDGE: Conservation Priorities Based on Threat and Phylogeny." *PLoS One* 2: e29.

Laurance, William F., Miriam Goosem, and Susan G. W. Laurance. 2009. "Impacts of Roads and Linear Clearings on Tropical Forests." *Trends in Ecology and Evolution* 24 (12): 659–69.

Megevand, Carole, with Aline Mosnier, Joël Hourticq, Klas Sanders, Nina Doetinchem, and Charlotte Streck. 2013. *Deforestation Trends in the Congo Basin: Reconciling Economic Growth and Forest Protection*. Washington, DC: World Bank.

Mooers, Arne Ø., Daniel Faith, and Wayne Maddison. 2008. "Converting Endangered Species Categories to Probabilities of Extinction for Phylogenetic Conservation Prioritization." *PLoS ONE* 3 (11): e3700. doi:10.1371/journal.pone.0003700.

Raleigh, Clionadh, Andrew Linke, Håvard Hegre, and Joakim Karlsen. 2010. "Introducing ACLED: An Armed Conflict Location and Event Dataset Special Data Feature." *Journal of Peace Research* 47 (5): 651–60.

Reymondin, Louis, Karolina Argote, Andy Jarvis, Carolina Navarrete, Alejandro Coca, Denny Grossman, Alberto Villalba, and Paul Suding. 2013. "Road Impact Assessment Using Remote Sensing Methodology for Monitoring Land-Use Change in Latin America: Results of Five Case Studies." Technical Note No. IDB - TN - 561, Inter-American Development Bank Environmental Safeguards Unit.

Uchida, Hirotsugu, and Andrew Nelson. 2009. "Agglomeration Index: Towards a New Measure of Urban Concentration." Background Paper for *World Development Report 2009*, World Bank, Washington, DC.

Wheeler, David, Dan Hammer, Robin Kraft, Susmita Dasgupta, and Brian Blankespoor. 2013. "Economic Dynamics and Forest Clearing: A Spatial Econometric Analysis for Indonesia." *Ecological Economics* 85 (January): 85–96.

Chapter 6

Wrapping Up: Main Messages, Caveats, and the Road Forward

Main Messages: What This Report Does

This report develops a new analytical tool aimed at better estimating the potential benefits of road improvements and the resulting decrease in transport costs, as well as the potential for negative environmental or conflict-related impacts. Although the costs of road improvement (for example, paving or expansion) are well understood by engineers, it is these other direct and indirect impacts that are especially challenging to identify. This challenge stems from the difficulty of obtaining reliable transport cost data and from the endogeneity of road placement. This report uses the most accurate transport data available and applies rigorous econometric analysis to arrive at more reliable estimates of the impacts of roads on a variety of welfare measures. Using these estimated elasticities, the report simulates the expected economic benefits stemming from alternative New Economic Partnership for Africa's Development (NEPAD) road improvement proposals. The report focuses on Nigeria and the Democratic Republic of Congo (DRC) as case studies, but depending on availability of data, the approach can be replicated in other countries. Chapters 2 and 3 consider the positive benefits that can result from road improvements, while chapters 4 and 5 present cautionary tales by considering conflict and the environment.

When evaluating the impact of roads on welfare, there are three potential sources of bias to consider: (1) spatial sorting by households, (2) endogenous emergence of markets, and (3) the nonrandom placement of roads connecting the two. Failure to take these factors into account could yield biased results. Throughout the analysis in this report, the first two sources of endogeneity are tackled through the inclusion of carefully chosen controls (in particular, household characteristics and marketshed fixed effects). The third source of bias, placement of roads, is addressed through an instrumental variable (IV) approach. One of the IVs most widely used in the literature as a means to correct for the nonrandom placement of roads is Euclidean distance. Though popular, this IV has come under criticism for its failure to take into account the physical obstacles along the path (mountains or lakes, for instance). In areas with irregular topography, this IV would potentially be weak because it would provide a less precise approximation of actual travel costs. To improve on the efficiency of the estimates, this report focuses on an alternative IV, termed the "natural path," which is the time it would take to walk from the enumeration area to the nearest market given the topography of the land and absent any roads. This natural path IV is used in chapters 2 and 3. In chapter 4, the natural path calculation is augmented with the inclusion of historical caravan routes in the DRC. Estimates using Euclidean distance are also generated, with results displayed in the chapter appendixes. Throughout, the estimates obtained are robust to the use of the Euclidean distance IV, alternative specifications, and relaxation of the exclusion restriction (following Conley, Hansen, and Rossi [2012]).

Chapter 2 launches the analysis with a look at how transport cost affects several indicators of welfare, with a focus on Nigeria. The outcome variables considered are household income (from either crop sales, livestock sales, or nonagricultural sources); probability of employment; asset wealth; the probability of being multidimensionally poor; and local GDP. By considering income, the analysis shows how roads affect short-term, flow variables. The results indicate that reducing transport cost to market by 10 percent increases crop revenue by 3.8 percent, livestock sales by 3.0 percent, and nonagricultural income by 3.9 percent. Considering the probability of employment (year-round employment and agriculture in particular) indicates that reduced transport cost induces a switch from agriculture to nonagricultural employment by workers. Furthermore, a 10 percent reduction in transport costs increases the asset wealth index by 2.0 percent and reduces the probability of the household being multidimensionally poor by 2.6 percent. Finally, the analysis broadens from the household to the surrounding area and estimates the elasticity of transport costs relative to local GDP (constructed by satellite nighttime light data from Ghosh et al. [2010]).

Using the elasticity of local GDP, which is spatially representative, this report simulates the impact of improving three proposed NEPAD projects: (1) a north-south corridor connecting Niger and Lagos (1,120 kilometers); (2) a northeastern corridor connecting Niger and Cameroon (940 kilometers); and (3) a southeastern corridor connecting Lagos and Cameroon (730 kilometers). A different elasticity was calculated for each of Nigeria's six geopolitical regions, so as to account for potential spatial heterogeneity. Benefits estimated by this methodology will accrue every year as long as the benefits from reducing transport costs and baseline GDP levels both remain constant.[1] To briefly recap the simulations, (1) the elasticity of local GDP relative to transport costs is estimated by instrumental variable (IV), (2) the estimated impact to local GDP is summed over the entire sample area to arrive at an aggregate benefit, (3) the estimated benefit to local GDP from each road project is compared to determine which had the largest impact per kilometer of road improved. The results suggest that improving the north-south corridor would yield the largest benefits ($970,000 per kilometer improved).

A reduction in transport costs would lead to an increase in welfare through many mechanisms, as reported in chapter 2. Chapter 3 extends the analysis by examining one such mechanism, the adoption of more modern technology in agriculture.[2] The chapter finds that technology and connectivity are linked, and should be treated jointly. Reducing transport costs increases production of crops grown using modern farming techniques, but has either a negative or negligible impact on production of traditionally farmed crops. This finding suggests that reduced transport cost leads to a switch from traditional to modern farming techniques. This result is reinforced by analysis using direct, survey-based observations of whether households report using a more modern farming technique, specifically mechanization, with the finding that reducing transport costs increases the probability of use.[3] Furthermore, while both nonmechanized and mechanized farmers see higher crop revenues from lower transport costs, the benefit to mechanized farmers is significantly higher. Taken together, these findings support the hypothesis that reducing transport costs would (1) induce higher adoption of modern technologies and (2) have a larger impact on modern farmers.

Whereas the first half of the report discusses the potential benefits stemming from road improvement, chapters 4 and 5 consider the potential negative consequences of roads. Much of Sub-Saharan Africa, especially the DRC, is plagued by chronic conflict. Chapter 4 addresses the question of whether road construction is still beneficial to welfare in a situation of ongoing violence. The results show that in the case of high levels of conflict, reducing transport costs leads to a fall in asset wealth of households as well as local GDP, and multidimensional poverty is seen to increase. These findings are consistent with the idea that better roads facilitate the movement of rebels. This study further finds a differential impact of transport costs depending on the location of conflict. Conflict near household has a significant positive effect on multidimensional poverty while conflict near market does not. This is possibly because when there is conflict near markets, households can at least retreat to subsistence agriculture, while they might not even be able to do that when there is conflict near households. In contrast, when conflict is relatively low, a reduction in transport costs leads to increased wealth and local GDP and reduced probability of multidimensional poverty. Thus, the main discovery is that the climate of conflict matters a great deal when deciding whether and when to invest in improving the roads. The results suggest that it would be prudent to wait for at least a reduction in the degree of fighting before launching a road improvement project.

Finally, chapter 5 examines the potential environmental costs of improving roads. Using novel high-resolution spatial data for eight Congo Basin countries, the study examines the impact of road improvement on deforestation and loss of biodiversity, a neglected area of research. Using the estimated elasticities of deforestation with respect to travel cost, the impacts of road improvement are simulated for eastern DRC, in much the same way as was done in chapter 2 using local GDP. Potential ecological costs should be considered early in the process of planning road improvements rather than after the road location has been determined.

Caveats

Although the approaches developed and used in this report arguably represent a clear advance in state-of-the-art research design, it is important to understand the limitations.

First, the study only focuses on two countries (Nigeria and the DRC), and in chapter 5 on deforestation in only one region (the Congo Basin). For that reason, the numerical results should not be used in other countries. The analysis would need to be carried out using the study's general methodological approach, but using data and assumptions appropriate to each country. Note, however, that very similar general results were found for local GDP—with, of course, different elasticities—for Burkina Faso, Niger, and Senegal, suggesting overall consistency in results across countries.

One obvious shortcoming is that the methodologies of the report cannot by themselves be the basis for making decisions. The approach does not consider costs of the road investments, so cannot be used alone as a benefit-cost analysis for establishing priorities in road construction, but rather just to inform the estimate of benefits. For the same reason, this approach does not allow road investments to be compared to alternative uses of public funds. Intersectoral tradeoffs with other kinds of investments—for example, in research, education, rural electrification—may, of course, be essential when considering how to get the biggest bang for the buck in rural development. Nor are multimodal transport investments modeled. Railroad connectivity is typically not economical for Africa given the low density of GDP, unless the output of a mine or some other high-volume commodity is to be transported. However, for the DRC water transport is potentially important (if it can be restored), but this issue was not explored in this report because of a lack of data.

An additional caveat is that realization of the benefits of road construction depends crucially on the presence of enabling factors. These factors include peace and security, but also other conditions such as sound governance—in fact, anything that contributes to a good business environment.

The benefit estimates from this methodology are more akin to partial than to general equilibrium impacts, given that macroeconomic linkages and second-order effects are not explicitly modeled. However, though not a perfect substitute for a structural model, the estimates produced using local GDP arguably proxy some general equilibrium effects if it is assumed that the empirical model is a reduced-form estimate of local GDP that correctly captures the true data-generation process. Thus, the methodology used here may capture effects that are indeed general equilibrium, but only in the area proximate to the road, not economywide.

Another potential shortcoming of this approach stems from the divergence between transport costs and transport prices in Africa. Part of the divergence results from monopolistic competition in the trucking sector, so it is not clear how much of a reduction in transport costs would translate into a reduction in tariffs to consumers of transport services. Although this may be a problem, the presence of an oligopoly does not imply that cost reductions are not passed at all to consumers, but rather that the reduction (pass-through) is less than dollar for dollar, as would be the case in a perfectly competitive market. Furthermore, the prices used in the calculation of transport costs were tested against real prices in Nigeria obtained from a survey and found to be similar, with no evidence of significant systematic divergence or markups.

A related issue in the context of access to markets for agriculture products is the intermediation through state-owned enterprises ("boards"), meaning that improvements in transport may not necessarily result in improvements in market access or better prices for farmers. For example, if there were a single buyer of a particular cash crop, then farmers would not necessarily enjoy an increase in farmgate prices for their produce corresponding to the full fall in transport costs. In short, market structure and the familiar double marginalization of vertical oligopolies (that is, there are sequential firms with market power) complicate the correlation between transport costs and prices. In practice, however, some of these distortionary structures have been lessened and many countries have abandoned marketing boards, including Nigeria, which has no remaining monopsony market board. So, although this is not an issue for this report's analysis using data from Nigeria, it could be an issue in other countries, and there could be other distortions that warrant consideration.

The Way Forward: Suggestions for Future Research

As noted, the analysis in this report was designed to capture direct household and local effects. Of course, some transport investment projects may have macroeconomic and economywide impacts that are more than marginal. Adequately capturing the general equilibrium effects resulting from the reduction in transport costs would require estimating a structural model of the entire economy, with accompanying data requirements. Such an exercise is certainly worth doing, but is beyond the scope of this report and is left for future research.

The analysis was designed to estimate benefits to average welfare (for example, average consumption, incomes, wealth, local GDP) and to multidimensional poverty, rather than looking more comprehensively at distributional effects. Future research could focus more explicitly on poverty, looking at the effects of reduced rural transport costs across the income distribution and considering some of the factors that condition whether individual households do indeed benefit directly from better market access (for example, education, access to credit).

It would also be useful for future research to model agglomeration effects. If there were more urban centers (populations of at least 50,000) closer to larger urban centers (such as around Lagos), one would expect larger agglomeration effects emanating from these large urban centers, and these could be explicitly incorporated into the analysis.

Notes

1. This would be a dubious assumption in the long term, but might be a reliable approximation over a short, three-to-five-year period.
2. There are, of course, other possible mechanisms, such as greater access to credit, but this is left for future research.
3. Although plenty of examples of modern farming techniques come to mind (irrigation, fertilizer, high-yielding seed varieties, to name a few) focusing on machinery provides the cleanest estimates because farmers' equipment choices are not as influenced by subsidies from the government, as is often the case with other modern technologies such as fertilizer and improved seeds.

References

Conley, Timothy G., Christian B. Hansen, and Peter E. Rossi. 2012. "Plausibly Exogenous." *Review of Economics and Statistics* 94 (1): 260–72.

Ghosh, T., R. L. Powell, C. D. Elvidge, K. E. Baugh, P. C. Sutton, and S. Anderson. 2010. "Shedding Light on the Global Distribution of Economic Activity." *Open Geography Journal* 3 (1): 148–61.

Geospatial Analysis

Introduction

Geospatial analysis is the application of statistical analysis and other informational techniques to data that have a geographical or geospatial aspect. Such analytical tools and associated modeling techniques typically use software capable of not only representing, mapping, and processing geodata, but also applying analytical methods, including the use of geographic information systems (GIS). The term "GIS" was first coined by the English geographer Roger Tomlinson, who introduced the concept of geographic information systems to the Canada Land Inventory in 1962.[1] Today the term is used to define a large field that provides a variety of capabilities designed to capture, store, manipulate, analyze, manage, and present all types of geographical data, and uses geospatial analysis in a variety of contexts, operations, and applications (Foote and Lynch 2014).

The basis for geospatial analysis can be traced back to the beginning of the twentieth century. For instance, centrographic statistics, the most basic type of descriptors of the spatial distribution of incidents, first appeared in 1926 with Lefever's pioneering work to measure geographic concentration (Levine 2013). Other examples of early geospatial analyses are the first statistical indicator of spatial autocorrelation produced by Moran (1950) and one of the oldest distance statistics, the nearest neighbor index, developed by Clark and Evans (1954), primarily for field work in botany.

Since the mid-1980s, GIS technology has rapidly evolved into practical applications for the public and private sectors. Technological advances in hardware (for instance, the launch of the Landsat satellite in 1973); better, cheaper, and more accessible computing power (the first mass-marketed personal computers emerged in the 1980s); the development of GIS software packages (ArcInfo and Grass in the 1980s); and the recent explosion of location-based data due to the ubiquity of mobile devices and the Internet (Google Earth was launched in 2005) have been the main drivers for this rapid evolution. Nevertheless, the ultimate success of GIS has been fundamentally driven by its applications in solving real-world problems.

In this report, GIS tools are used extensively for various objectives, from simple extraction of spatial data to performing advanced spatial analyses that create data never used before. This section presents only a selection of the most advanced geospatial analyses that are used. Specifically, the appendix describes the process used to create four important data sets used in the analyses in this book: transport costs to market, the natural-historical path, conflict kernels, and a social fractionalization index.

Road Network GIS Data Set

The costs of traveling between any two points along a road network depend upon on a number of factors—distance, road conditions, terrain, and type of vehicle are some of the most important. To build a data set capable of answering the question "how much does it cost to go from point A to point B?" for every location in a certain country, a network data set must be constructed. Network data sets are the basic structures for modeling transportation networks. They are created from network elements that can include lines (edges), points (junctions), and turns, and store the connectivity of the source features.[2] A key element of the network data set is the network attributes, which can be defined as the properties of the network elements that control navigation over the network. Examples of attributes include the time needed to travel a given length of road, which streets are restricted for which vehicles, speeds along a given road, and one-way streets.

A network data set was not available for either Nigeria or the Democratic Republic of Congo (DRC). In fact, the challenge was even larger: no suitable and comprehensive vector data set with a good geometric representation of primary and rural roads or road attributes was available.[3] The following section describes how this shortcoming was overcome in Nigeria. A similar procedure was followed to obtain transport cost to market in the DRC, but it is not described here because of space considerations.

Nigeria's Road Network Data Set Creation
The foundation of a good network data set is a complete vector data set. The last available GIS data set for Nigeria was from 1998 and only contained primary roads; thus, the need to create a new one from scratch was unquestionable. After careful evaluation of open source data sets (Google Maps, Open Street Maps, and VMAP0) it was determined that Nigeria's rural roads were not represented properly in any freely available source. The best alternative found was a licensed data set from DeLorme's Digital Atlas of the Earth 2012 in which the roads (primary, secondary, and tertiary) are drawn from Landsat imagery and have excellent coverage.[4] The downside of DeLorme's road data is that it only includes

functional classes (primary, secondary, rural, urban, and so on) but lacks road quality and surface type as valid attributes for estimating the cost of traveling through each segment.

To resolve the attribute data limitation, a dual approach was followed. For primary roads, a visual inspection data survey was acquired from Nigeria's Federal Roads Maintenance Agency (FERMA) for 2007 and 2010. The survey was georeferenced by linking every road record to a vector line in DeLorme's road network shapefile.

For secondary and rural roads, a specialized road survey instrument was developed and deployed in each of Nigeria's 36 states, using the local offices of the World Bank's Fadama project. The process of collecting the crucial road data and building a countrywide data set to assess market connectivity involved generating a list of the key road segments by state that capture connectivity between larger markets as well as a sample of secondary roads representing rural access. The survey instrument consisted of a set of maps for each state identifying road segments of interest and a spreadsheet to record responses.[5] The resulting data were processed, cleaned, and georeferenced into DeLorme's road network shapefile. Finally, the FERMA and Fadama results were merged into a complete road network that was later validated in Nigeria by transport specialists from the government, the World Bank, and the United Kingdom's Department for International Development.

The final input needed to estimate the cost of transport is terrain. The process of adding a topography attribute into the network data set consisted of calculating an average slope for each road segment. The raw topographic data used was the Hydrologically Corrected/Adjusted Shuttle Radar Topography Mission Digital Elevation Model with 90 meter resolution (Vagen 2010). Finally, the slopes were translated into relief classes following the United Nations Food and Agriculture Organization's classifications: Flat (0–2 percent), Undulating (2–8 percent), Rolling (8–16 percent), Hilly (16–30 percent), and Mountainous (> 30 percent).

The next step was to estimate the monetary transport cost for each type of road segment.

Obtaining Transport Cost from HDM-4

To translate road attributes into a monetary cost, the study used the Highway Development and Management Model (HDM-4). HDM-4 is a software system for evaluating options for investing in road transport infrastructure and estimating the social benefits as measured by reduced costs to road users. Pecuniary road user costs are characterized for a given country as a function of the vehicle fleet unit costs, utilization rate, and road characteristics. The assumptions used in HDM-4 in Nigeria and the DRC follow.

Characterization of Network Type and Terrain

The road networks of Nigeria and the DRC include three classes of roads: primary, secondary, and tertiary. The average vehicle speed and width of the main carriage road were used to characterize the differences among network types as follows:

Paved Road Speed (km/hour) by Network and Condition for Flat Terrain

Road condition	Primary 7m	Secondary 6m	Tertiary 5m
Good	80.0	65.0	50.0
Fair	68.0	55.3	42.5
Poor	48.0	39.0	30.0

Note: m = meters (road width).

Unpaved Road Speed (km/hour) by Network and Condition for Flat Terrain

Road condition	Primary 7m	Secondary 6m	Tertiary 5m
Good	64.0	52.0	40.0
Fair	54.4	44.2	34.0
Poor	38.4	31.2	24.0

Note: m = meters (road width).

Paved Road Speed (km/hour) by Network and Condition for Rolling/Mountainous Terrain

Road condition	Primary	Secondary	Tertiary
Good	72.0	58.5	45.0
Fair	57.6	46.8	36.0
Poor	36.0	29.2	22.5

Unpaved Road Speed (km/hour) by Network and Condition for Rolling/Mountainous Terrain

Road condition	Primary	Secondary	Tertiary
Good	57.6	46.8	36.0
Fair	46.1	37.4	28.8
Poor	28.8	23.4	18.0

Terrain type is defined using the following concepts and road geometry parameters:

• Flat: Mostly straight and gently undulating

• Rolling: Bendy and gently undulating

• Mountainous: Winding and gently undulating

Characterization of Network Type and Condition

The International Roughness Index (IRI) (meters/kilometer) was used to define the differences in road condition by network as follows:

Paved Road IRI (m/km) by Network and Condition

Road condition	Primary 7m	Secondary 6m	Tertiary 5m
Good	2	3	4
Fair	5	6	7
Poor	8	9	10

Note: m = meters (road width); m/km = meters per kilometer.

Unpaved Road IRI (m/km) by Network and Condition

Road condition	Primary 7m	Secondary 6m	Tertiary 5m
Good	6	8	10
Fair	12	13	14
Poor	16	18	20

Note: m = meters (road width); m/km = meters per kilometer.

Characterization of Vehicle Type and Estimation

A heavy truck was defined as the typical vehicle for modeling freight transport costs and it was assumed that the truck can transport an average weight of 15 tons (average net weight). For each country, locally calibrated input data were gathered for the value of used vehicle, tire, and fuel cost, maintenance labor cost, and driver cost, among others.

Cost per Ton per Kilometer for Nigeria for Different Types of Roads

Finally, using the parameters above, a final cost per ton-kilometer for each road type was estimated ($/ton/km):

Paved Flat ($/ton/km) by Network and Condition

Road condition	Primary	Secondary	Tertiary
Good	0.0526	0.0529	0.0533
Fair	0.0570	0.0583	0.0596
Poor	0.0617	0.0637	0.0986

Paved Rolling ($/ton/km) by Network and Condition

Road condition	Primary	Secondary	Tertiary
Good	0.0533	0.0531	0.0535
Fair	0.0577	0.0586	0.0599
Poor	0.0623	0.0643	0.0996

Paved Mountainous ($/ton/km) by Network and Condition

Road condition	Primary	Secondary	Tertiary
Good	0.0574	0.0562	0.0584
Fair	0.0620	0.0615	0.0644
Poor	0.0675	0.0676	0.1055

Unpaved Flat ($/ton/km) by Network and Condition

Road condition	Primary	Secondary	Tertiary
Good	0.0629	0.0673	0.0730
Fair	0.0795	0.0831	0.0867
Poor	0.0941	0.1017	0.1091

Unpaved Rolling ($/ton/km) by Network and Condition

Road condition	Primary	Secondary	Tertiary
Good	0.0618	0.0678	0.0752
Fair	0.0801	0.0837	0.0877
Poor	0.0945	0.1021	0.1095

Unpaved Mountainous ($/ton/km) by Network and Condition

Road condition	Primary	Secondary	Tertiary
Good	0.0651	0.0748	0.0868
Fair	0.0820	0.0884	0.0974
Poor	0.0954	0.1038	0.1130

Cost per Ton per Kilometer for the DRC for Different Types of Roads

Paved Flat ($/ton/km) by Network and Condition

Road condition	Primary	Secondary	Tertiary
Good	0.1174	0.1192	0.1237
Fair	0.1226	0.1264	0.1293
Poor	0.1286	0.1299	0.1349

Paved Rolling ($/ton/km) by Network and Condition

Road condition	Primary	Secondary	Tertiary
Good	0.1190	0.1191	0.1231
Fair	0.1241	0.1268	0.1302
Poor	0.1305	0.1315	0.1367

Paved Mountainous ($/ton/km) by Network and Condition

Road condition	Primary	Secondary	Tertiary
Good	0.1283	0.1292	0.1312
Fair	0.1333	0.1318	0.1382
Poor	0.1410	0.1391	0.1449

Unpaved Flat ($/ton/km) by Network and Condition

Road condition	Primary	Secondary	Tertiary
Good	0.1401	0.1463	0.1559
Fair	0.1622	0.1755	0.1901
Poor	0.1976	0.2133	0.2290

Unpaved Rolling ($/ton/km) by Network and Condition

Road condition	Primary	Secondary	Tertiary
Good	0.1348	0.1453	0.1588
Fair	0.1638	0.1771	0.1921
Poor	0.1991	0.2147	0.2305

Unpaved Mountainous ($/ton/km) by Network and Condition

Road condition	Primary	Secondary	Tertiary
Good	0.1390	0.1570	0.1806
Fair	0.1681	0.1857	0.2091
Poor	0.2014	0.2186	0.2379

Transport Cost as an Impedance for Network Analysis

The last step was to add the individual transport cost for each combination of road types (54 in total) into the network data set segments and multiply them by the length of the roads (see map A.1). As a result, a monetary road-user cost could then be used as a measure of the amount of resistance required to traverse a path in a network, or to move from one element in the network to another. Higher impedance values indicate more resistance to movement; a value of zero indicates no resistance. Using this approach, an optimum path in a network is the path of lowest impedance, also called the least-cost path.

Finally, these costs per ton-kilometer were used to calculate the cost for each georeferenced household or pixel centroid to transport one ton of goods to the nearest market.

Map A.1 Construction of Road-User Cost Network Data Set

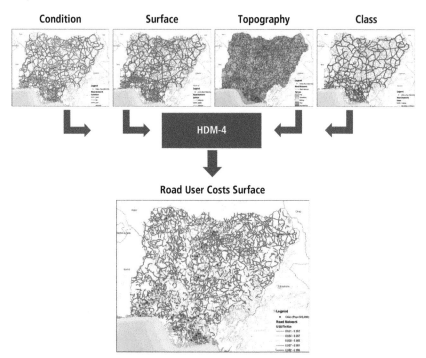

Sources: DeLorme; Nigeria Federal Roads Maintenance Agency; SRTM 90m (dataset), NASA; Monchuk et al. (2013); and calculations.
Note: HDM-4 = Highway Development and Management Model.

Natural-Historical Path for the DRC

Since the arrival of the Portuguese mariner Diego Cão in 1483, Congo (Kingdom of Kongo at that time) has had cultural, social, and economic connections with Europe. Western religions, literacy, the wheel, the plow, the gun, and many other technologies were quickly adopted by the Congolese (Acemoglu, Robinson, and Woren 2012). However, these came at great expense: one of the principally traded goods in exchange was slaves (British Museum 2011).

As contact with Europe deepened, other types of goods were introduced, such as ivory, rubber, copper, diamonds, raffia cloth, pottery, and other natural resources. Trade with Europe was based in the coastal cities of Sonyo and Pinda, so an extensive trade network was required to access the eastern part of the country (primarily near present day Kivu and Katanga Provinces) where much of the mineral deposits and other natural resources were mined. Fueled by the

Map A.2 DRC Historic Trade Routes circa 1896 and Digitally Manipulated Map with Routes

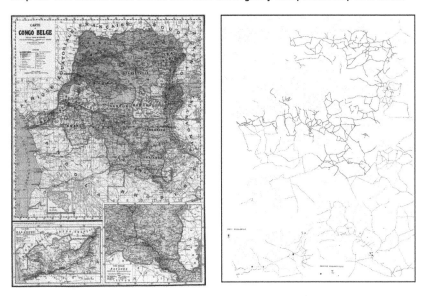

Source: Carte du Congo Belge (Moor 1896). Library of Congress, Geography and Map Division.
Note: DRC = Democratic Republic of Congo.

industrial revolution and new inventions, such as inflatable rubber tubes, the demand for goods increased dramatically (Schimmer 2010). By the end of the nineteenth century, the Congo was a personal possession of King Leopold II (not an official Belgian colony). The king was engaged in a vigorous publicity campaign aimed at convincing the other European powers to recognize the legitimacy of his rule, a difficult task in view of the notorious brutality of his administration in Africa. One of the products of King Leopold's "Office of Publicity" is a very detailed "Carte du Congo Belge" (Map of the Belgian Congo),[6] which includes caravan routes and existing and projected railways (map A.2). This map, which shows the transport network constructed to move slaves, ivory, and mineral resources between the interior of Congo and the coastal harbors, is the main input used to construct the natural-historical instrumental variable.

The natural-historical instrumental variable is constructed by merging two sources: the historical caravan route map described above, and a natural walking path map calculated for this study. The natural walking path, or natural path, is created by estimating the time-minimizing route a pedestrian would travel over land, absent the benefits of a road network. The next section details how the natural path network was calculated.

Natural Path Walking Time Calculation

The first step was to create a GIS cost-surface model that accounts for all the traffic off-road or outside the caravan network for 1900. To do so, this report followed an approach similar to that used to construct a global map of accessibility in the World Bank's *World Development Report 2009: Reshaping Economic Geography* (Uchida and Nelson 2009). The surface model, or off-path friction-surface raster,[7] is a grid in which each pixel contains the estimated time required to cross that pixel by walking. To create this raster, two basic layers were combined: terrain slope and land cover.

The slope raster was calculated from NASA's Shuttle Radar Topography Mission (SRTM)[8] Digital Elevation Models (DEMs) with a resolution of 90 meters. Even though the original topography data were obtained in February 2000, it was assumed that there has not been any drastic change in the DRC's terrain in the twentieth century. Under this assumption, the SRTM 90-meter data set provides a fairly good representation of the elevation terrain circa 1900.

Land cover data are far more challenging. With the surge of remote sensing technology, several land cover and land use data sets have been created in the past few years. These high-resolution data sets are very accurate representations of the current state of physical material at the surface of the earth. However, these data sets cannot be used in this analysis because they are not a good representation of the land cover for 1900. Land has changed rapidly in the last hundred years in the DRC: deforestation, open pit mining, and urbanization have drastically transformed the surface. Therefore, other data sets were used. The Oak Ridge National Laboratory (ORNL)[9] has developed a Historical Land Cover and Land Use data estimate (Goldewijk 2010). This data set describes historical land use changes over a 300-year historical period (1700–1990) and was modeled based on a deep understanding of the global environment, historical statistical inventories of agricultural land (census data, tax records, land surveys, and the like), and different spatial analysis techniques. A shortcoming of this data set is that the resolution is approximately 55 kilometers per cell, making a clear tradeoff between space (resolution) and time (representative for 1900). However, given the importance of obtaining an accurate picture of the historical land cover, the ORNL data set, despite its low resolution, was chosen over the newer, better resolution data because it provides a better account of land surface types. Map A.3 displays the terrain roughness (left) and land cover (right) for the DRC, circa 1900.

The off-path friction-surface raster was created by combining the land topography raster and the land cover map. This approach was based on the guiding assumption that all travel in 1900 was on foot, and walking speed is therefore determined by the land cover class and slope. The typical velocity of a hiker when walking on uneven or unstable terrain is 4 kilometers per hour,

Map A.3 Terrain Roughness and Land Cover, circa 1900

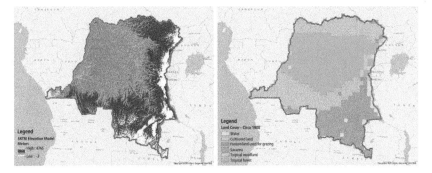

Source: Calculations using SRTM 90m (dataset), NASA; ISLSCP II (International Satellite Land-Surface Climatology Project, Initiative II) (dataset), ORNL DAAC.

which diminishes on steeper terrain. A hiking velocity equation (Tobler 1993) was used to reflect changes in travel speed as a function of trail slope:

$$W = 6 \times \exp(-3.5 \times |S + 0.05|)$$

in which W is the hiking velocity in kilometers per hour and S is the slope or gradient.

By applying the speed formula, the time required to cross 1 pixel (92.5 meters) was computed. In this way, the time (hours) that it takes to walk through any pixel— only taking into account the topography—was calculated, as shown in map A.4. Note that the more mountainous regions of the DRC near the Kivu provinces in the east have significantly greater walking times.

Next, a delay factor to account for the effect of walking through different land classes was estimated. The historical land cover raster resolution was changed from a half degree to 90 meters and each class was assigned a speed-reducing factor according to table A.1.

Finally, walking travel speed (the slope variable) was multiplied by the delay factor (the land cover variable) to obtain the off-path friction-surface raster that models the time it takes to walk 92.5 meters anywhere in the DRC circa 1900 (map A.5).

Historical Path Creation

The second step was to digitize a historical map of the DRC to create a shapefile that could be added as a layer for spatial analysis. A mix of image manipulator (open source GIMP [http://www.gimp.org/]) and GIS software (ESRI's ArcGis Desktop) was used to separate the routes from other map features and then digitize the map. Map A.2 shows the historical map, which was then converted into the shapefile shown in map A.6.

Map A.4 Walking Time in the DRC Given Land Topography (hours per pixel)

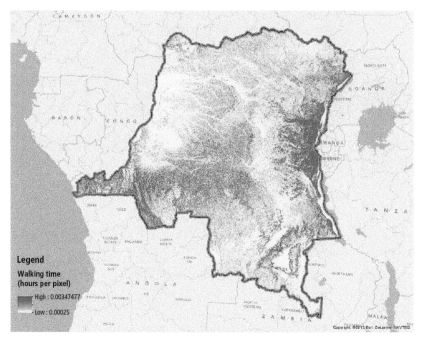

Legend
Walking time
(hours per pixel)
High : 0.00347477
Low : 0.00025

Source: SRTM 90m (dataset), NASA; ISLSCP II (dataset), ORNL DAAC.
Note: DRC = Democratic Republic of Congo.

Table A.1 Speed-Reducing Factors

Class #	Biome Type	Delay Factor
0	Oceans or water	n.a.
1	Cultivated land	1.00
2	Pasture or land used for grazing	1.00
5	Ice	1.33
6	Tundra	1.00
7	Wooded tundra	1.00
8	Boreal forest	1.17
9	Cool conifer forest	1.00
10	Temperate mixed forest	1.17
11	Temperate deciduous forest	1.33
12	Warm mixed forest	1.17
13	Grassland or steppe	1.00
14	Hot desert	1.00

(continued next page)

Table A.1 (continued)

Class #	Biome Type	Delay Factor
15	Scrubland	1.00
16	Savanna	1.00
17	Tropical woodland	1.33
18	Tropical forest	1.67
19	No data over land (for example, Antarctica)	n.a.

Source: Calculations based on Uchida and Nelson (2009).
Note: n.a. = not applicable.

Map A.5 Final Natural-Historical Path Raster (hours per pixel)

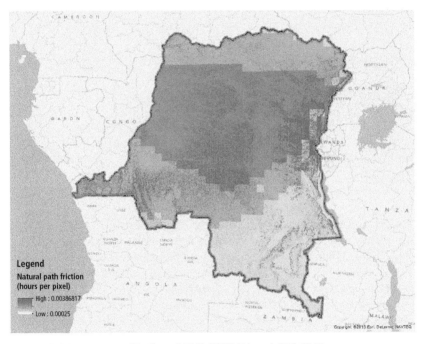

Source: Calculations using SRTM 90m (dataset), NASA; ISLSCP II (dataset), ORNL DAAC.
Note: DRC = Democratic Republic of Congo.

Then the newly created shapefile was converted into a raster with a resolution of 92.5 meters to match that of the natural path friction-surface raster. The pixel value assigned to every cell through which a caravan route passes is approximately 0.02 hours or 1.2 minutes. This value was arrived at by assuming the caravan route travel speed was 5 kilometers per hour,[10] which is equivalent to the average human walking speed on stable terrain.

Map A.6 Historical Caravan Route Shapefile

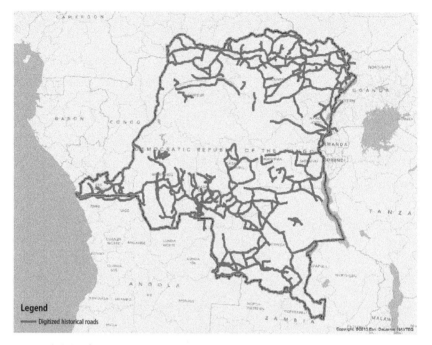

Source: Calculations from Carte du Congo Belge (Moor 1896).

Cost-Distance Function: Calculating Travel Time

The third step was to merge the off-path and the caravan route friction surfaces. ArcGIS for Desktop's tool MERGE was used to combine the two rasters into a single one in which the order of the input defines the order of precedence, in this case the caravan routes overlay off-path walking. Then the friction surface was obtained to model the time it takes to move around the entirety of the country in approximately 1900, taking into account terrain, land cover type, and transport infrastructure.

To create the final variable, which was used as an instrumental variable in this study, the time that it takes to travel on foot from each pixel in the study area to different selected cities or target destinations was estimated. ArcGIS Cost distance tools were used to calculate, for each pixel, the least cumulative amount of walking time between a pixel and a specified location (market). The algorithm uses the node/link cell representation, whereby the center of each cell is considered a node and each node is connected to its adjacent nodes by multiple links. Every link has an impedance derived from the costs (measured in units of time) associated with the cells from the natural path friction-cost

Map A.7 Travel Time to Kinshasa in 1900

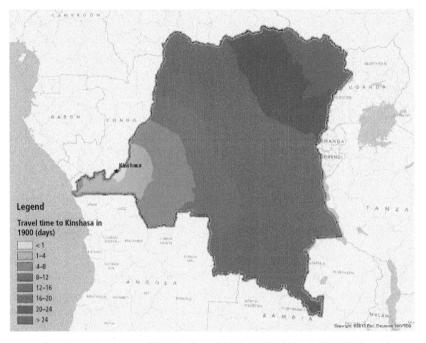

Sources: Calculations using data from SRTM 90m (dataset), NASA; ISLSCP II (dataset), ORNL DAAC; and Carte du Congo Belge (Moor 1896).

surface and from the direction of movement through the cells. See map A.7 for an example of a raster measuring travel time from each point to Kinshasa.

The creation of a least-cumulative-cost raster was replicated for each of the DRC's 57 selected cities, then the Spatial Production Allocation Model locations (specified as a point feature, see chapter 3) were overlaid on top of the travel time rasters and finally the cell values were retrieved. An origin-to-destination travel time matrix of 27,500 rows (equal to the number of pixels in the DRC) and 57 columns (the selected cities) was thus obtained. Finally, this data set was compared with the current travel cost data set, and the appropriate city for each pixel was selected for the econometric model.

Conflict Kernel (DRC)

Researchers studying conflict have increasingly embraced a micro-level approach as an alternative to analyses at the macro, country-year level. The micro-level approach suggests that researchers should focus on subnational or individual-conflict units of analysis, which are better suited to the study of dynamics across time and space, better enabling inferences about the local conditions that affect, and are affected by, conflict to be drawn. Within the drive for empirical disaggregation, two trends have emerged: the usual approach of using household or individual surveys and the newer one of geographical and temporal disaggregation of conflict events. The latter approach was selected for this report. Specifically, the Armed Conflict Location Events Dataset (ACLED) was used.

The ACLED project codes report information on the location, date, and other characteristics of politically violent events in unstable and warring states. ACLED focuses specifically on tracking rebel, militia, and government activity over time and space; locating rebel group bases, headquarters, strongholds, and presence; distinguishing between territorial transfers of military control from governments to rebel groups and vice versa; recording violent acts between militias; collecting information on rioting and protesting; and documenting nonviolent events that are crucial to the dynamics of political violence (for example, rallies, recruitment drives, peace talks, high-level arrests). ACLED Version 4 data cover all countries on the African continent from 1997 to 2013.

However, ACLED could not be used in its raw format because technical and methodological problems arise based on the nature of conflict. First, using points pins the conflict to a specific location and does not capture the effects of conflict on the surrounding area (for instance, battles that may have been fought in a large area). Second, conflict points cannot capture conflict intensity very well (for instance, one isolated conflict point versus a cluster of conflict points). Finally, ACLED is subject to some geographic imprecision based on how the data were obtained (for instance, rural conflicts are often allocated to nearby villages).[11]

To account for these methodological issues, a "hot spot" strategy was followed. Hot spots are concentrations of incidents within a limited geographic area that appear over time (Braga and Weisburd 2010). In particular, a kernel density interpolation technique was chosen because it allows conflict points to be transformed into a smooth surface, thus generalizing conflict locations. To calculate the value at any point, the kernel density function takes a weighted average of all the conflicts around that point to create the surface. The magnitude of the weight declines with distance from the point according to the chosen kernel function.[12]

When using kernel density estimation, two decisions must be made: what kernel function to use, and what bandwidth to search over. The literature using this technique in the context of conflict is not very large, and therefore does not offer much guidance on these decisions. Although other researchers have used similar techniques to estimate crime densities (for example, Levine 2006; Chainey, Tompson, and Uhlig 2008; Eck et al. 2005), there appears to be no agreement on the most appropriate kernel function. Without an obvious candidate, a quadratic kernel function, as described in Silverman (1986, 76), was used for this analysis. This is a fairly common kernel function. Other kernel functions used in its stead (such as Gaussian, Epanechnikov, quartic) resulted in no large effects on the kernel surface.

The other key parameter in the kernel function is the bandwidth. Without any robust theory describing how far the effects of conflict can permeate, this study relies on spatial autocorrelation. Using the Moran correlogram (Levine 2013), the degree of spatial autocorrelation as a function of distance was calculated. By measuring spatial autocorrelation for a series of distances and their corresponding z-scores, a sense of the intensity of spatial clustering can be obtained. Statistically significant peak z-scores will show distances where spatial processes promoting clustering are most pronounced. Figure A.1 shows that the distance at which autocorrelation is most pronounced is approximately 50 kilometers, and this is the bandwidth employed in the kernel function.

The last ingredient needed for the analysis is the conflict data itself. As explained, the data not only include where and when the events happened, but also the type (violent, battles, riots, and so forth) and outcome (number of fatalities), among other variables. Therefore, several measures of conflict were estimated. The first was the "kernelly" estimated number of fatalities in the five

Figure A.1 Moran's I Statistic, Spatial Autocorrelation by Distance

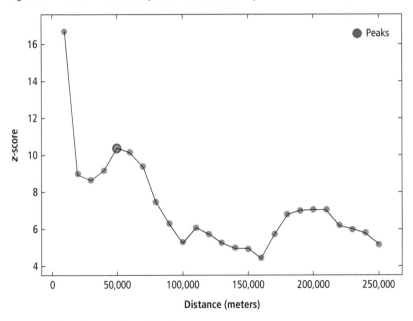

Source: Calculations from Raleigh et al. (2010).

Map A.8 Raw ACLED Conflict Data (left) versus Kernelly Estimated Conflict Raster (right)

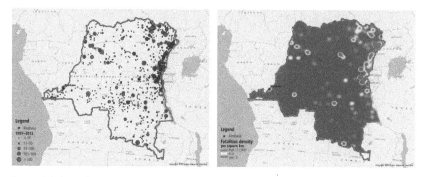

Source: Calculations from Raleigh et al. (2010).
Note: ACLED = Armed Conflict Location Events Dataset.

years preceding the Demographic and Health Survey data set (2003–07) and local GDP data set (2002–06). This variable was calculated around each household and also around each market. Also, a dummy variable that indicates if there were relatively high levels of conflict near households and markets was generated; it takes a value of 1 if the kernelly estimated fatalities are greater than the median number of fatalities due to violent conflict both near households and around the nearest market, and zero otherwise.

To illustrate how the conflict data were modified to facilitate analysis, map A.8 shows the original ACLED conflict data (left), as well as the kernel density map (right).

Social Fractionalization Index

As discussed in chapter 4, conflict is endogenous and is instrumented by a social fractionalization index. This section discusses how this variable is calculated. Social fractionalization, which measures religious, ethnic, and linguistic diversity within a country or region, can be used as a valid instrument for conflict because it can be a strong determinant of the probability, duration, and levels of conflict.[13] Typically, fractionalization indices are calculated using nationwide census data. However, such data are not available in the DRC. Instead, a spatial approach was used to calculate micro-level ethnic fractionalization.

The raw ethnicity data come from the magnificent *Peoples of Africa Atlas* by Felix and Meur (2001). Felix and Meur delimit the map by "ethnic group," defined as a large number of people with a sense of belonging together and who share a significant number of cultural criteria that make them feel different from neighboring groups. Such an endeavor comes with its fair share of difficulties. Time and partial boundaries of the maps are understandably fuzzy. The authors claim that "the picture" they portray has been taken in the "ethnographical past," but also recognize that the borders drawn varied before this snapshot was taken, and have continued to move or may have even disappeared since. However, this effort is perhaps the only coherent mapping study since the seminal work of Murdock (1959), and it is the outcome of more than 15 years of research and fieldwork using the most reliable and up-to-date sources.

To create a fractionalization index, a similar approach to that of the Herfindahl Index of ethnolinguistic group shares (Alesina et al. 2003) was followed. However, the number of people within a region from a particular ethnic group was replaced by the size of the land area occupied by a given ethnicity, assuming a uniform population density across different ethnicities. Formally, the calculation is given by

$$FRACT_w = 1 - \sum_K P_k^2, \gamma \in 50km,$$

in which $FRACT_w$ is the fractionalization index of location w (w can be a household or market) and p_k is the percentage of land in which ethnicity k is the dominant ethnic group, within the 50 kilometer bandwidth of location w,[14] where w can be a household or a market. Higher (lower) values of the fractionalization index imply higher (lower) levels of ethnic fractionalization. In the extreme case, only one group inhabits a region, thus $p_k = 1$ and $FRACT_w = 0$. Specifically, this measure gives the probability that any two points of land chosen at random will have different dominant ethnic groups.[15]

To estimate the percentage of land that each group occupies, a digitized version of the Felix and Meur atlas (Felix and Meur 2001) was obtained from the Harvard Center for Geographic Analysis's AfricaMap project.[16] For example, to estimate the fractionalization index around Kisangani, a circle or buffer of 50 kilometer radius was created, then the area that each ethnicity occupies within that circle was estimated to arrive at the percentage of land area that each tribe or ethnicity dominates. The values were input into the fractionalization index, and the index value of 0.83 was obtained (figure A.2). This implies that Kisangani has a very high level of ethnic fractionalization.

This exercise was replicated using a geographic information system script to obtain the percentage of land occupied by different ethnicities around each selected study area (buffers of 50 kilometers around either cities of 50,000 people or more, Demographic and Health Survey households, or individual pixels for local GDP estimation). The 50 kilometer radius around each household was used as an instrument for conflict around each household, and the 50 kilometer radius around each market was used as an instrument for conflict around markets.

Figure A.2 Fractionalization Calculation for Kisangani

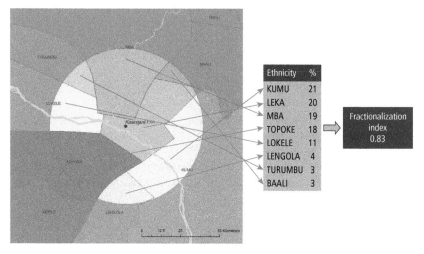

Source: Calculations from Felix and Meur (2001).

Notes

1. GIS Lounge website at http://www.gislounge.com/gis-timeline/.
2. Edges connect to other elements (junctions) and are the links over which agents travel; junctions connect edges and facilitate navigation from one edge to another; turns store information that can affect movement between two or more edges.
3. A vector data set is a coordinate-based data set representing geographic features as points, lines, and polygons.
4. DeLorme uses Landsat 7 to draw roads so the vectors fall within the recognized casing of the roadway. Landsat 7 provides the horizontal positional accuracy of +/− 50 meters, with a 90 percent confidence interval.
5. For more information about the survey and GIS methodology, see "Spatial Analysis and GIS Modeling to Promote Private Investment in Agricultural Processing Zones: Nigeria's Staple Crop Processing Zones," presented at the Annual World Bank Conference on Land and Poverty 2013.
6. Carte du Congo Belge / éditée par l'Office de Publicité, anciens établissements J. Lebègue & Cie. - Éditeurs, Bruxelles (1896). Stored at the Library of Congress and downloaded from http://www.wdl.org/en/item/59/.
7. A raster is a geographic map with information (for example, elevation, land cover) in a matrix of pixels.
8. Information about the SRTM is available at http://www2.jpl.nasa.gov/srtm/.
9. ORNL is a multiprogram science and technology laboratory managed for the U.S. Department of Energy by UT-Battelle, LLC.

10. https://en.wikipedia.org/wiki/Walking.

11. In ACLED, the geographic uncertainty level is coded with "geoprecision codes" ranging from 1 to 3 (higher numbers indicate broader geographic spans and thus greater uncertainty about where the event occurred). A geoprecision code of 1 indicates that the coordinates mark the exact location where the event took place. When a specific location is not provided, ACLED selects the provincial capital. ACLED may thus attribute violent incidents to towns when in fact they took place in rural areas, thereby introducing a systematic bias toward attributes associated with urban areas that can lead to invalid inferences.

12. For instance, if a conflict occurs exactly on the point that is being calculated, the value of that conflict will receive a weight of 1. A conflict that is 5 kilometers away from the point will receive a weight of α and a conflict 10 kilometers away will receive a weight of β, where $1 > \alpha > \beta > 0$. Eventually, at some distance, referred to as the bandwidth, the weight becomes zero.

13. Using a similar argument, Mauro (1995) uses ethnolinguistic fractionalization to instrument for institutional quality when estimating its effect on GDP growth.

14. The distance of 50 kilometers was chosen because that is the same radius as the bandwidth for the conflict kernels; therefore, conflict and fractionalization are measured within the same area.

15. This index traditionally gives the probability that any two persons chosen at random will be from different groups (ethnic, religious, linguistic, and so forth, depending on what the researcher is studying). This index is modified slightly to accommodate the fact that the ethnicity data do not observe individuals, but only land area. Rather than considering how many people of different ethnicities live in a certain area, the amount of land area that belongs to a plurality of each ethnicity is calculated. The number of people from a given ethnicity variable was then replaced in the fractionalization index with land area occupied by a given ethnicity.

16. http://worldmap.harvard.edu/data/geonode:Ethnicity_Africa_2001_Felix_Scan.

References

Acemoglu, Daron, James A. Robinson, and Dan Woren. 2012. *Why Nations Fail: The Origins of Power, Prosperity, and Poverty*. Vol. 4. New York: Crown Business.

Alesina, Alberto, Arnaud Devleeschauwer, William Easterly, Sergio Kurlat, and Romain Wacziarg. 2003. "Fractionalization." *Journal of Economic Growth* 8 (2): 155–94.

Braga, Anthony A., and David L. Weisburd. 2010. *Policing Problem Places: Crime Host Spots and Effective Prevention*. Oxford: Oxford University Press.

British Museum. 2011. "The Wealth of Africa, The Kingdom of Kongo." http://www .britishmuseum.org/pdf/KingdomOfKongo_TeachersNotes.pdf.

Chainey, Spencer, Lisa Tompson, and Sebastian Uhlig. 2008. "The Utility of Hotspot Mapping for Predicting Spatial Patterns of Crime." *Security Journal* 21 (1): 4–28.

Clark, Philip J., and Francis C. Evans. 1954. "Distance to the Nearest Neighbor as a Measure of Spatial Relationships in Populations." *Ecology* 35 (4): 445–53.

Eck, John E., Spencer Chainey, James G. Cameron, Michael Leitner, and Ronald E. Wilson. 2005. "Mapping Crime: Understanding Hotspots." National Institute of Justice, Washington, DC.

Felix, Marc L., and Charles Meur. 2001. *Peoples of Africa Atlas: An Enthnolinguistic Atlas.* Brussels: Congo Basin Art History Research Center.

Foote, Kenneth E., and Margaret Lynch. 2014. "The Geographer's Craft Project." Department of Geography, The University of Colorado at Boulder. http://www .colorado.edu/geography/gcraft/notes/intro/intro.html.

Goldewijk, K.K. 2010. "ISLSCP II [International Satellite Land-Surface Climatology Project, Initiative II] Historical Land Cover and Land Use, 1700–1990." In Hall, Forest G., G. Collatz, B. Meeson, S. Los, E. Brown de Colstoun, and D. Landis, eds. ISLSCP Initiative II Collection. Data set. Available online (http://daac.ornl.gov/) from Oak Ridge National Laboratory Distributed Active Archive Center, Oak Ridge, Tennessee, U.S.A. doi:10.3334/ORNLDAAC/967.

ISLSCP II (International Satellite Land-Surface Climatology Project, Initiative II) Historical Land Cover and Land Use, 1700–1990 (data set). Oak Ridge National Laboratory Distributed Active Archive Center (ORNL DAAC), National Aeronautics and Space Administration (U.S.) (NASA). https://daac.ornl.gov/ISLSCP_II/guides /historic_landcover_xdeg.html.

Levine, Ned. 2006. "Crime Mapping and the Crimestat Program." *Geographical Analysis* 38 (1): 41–56.

———. 2013. "CrimeStat: A Spatial Statistics Program for the Analysis of Crime Incident Locations." (v 4.0). Ned Levine and Associates, Houston, Texas; and the National Institute of Justice, Washington, DC.

Mauro, Paolo. 1995. "Corruption and Growth." *Quarterly Journal of Economics* 110 (3): 681–712.

Monchuk, Daniel, Alvaro Federico Barra, John Nash, and Siobhan Murray. 2013. "Spatial Analysis and GIS Modeling to Promote Private Investment in Agricultural Processing Zones: Nigeria's Staple Crop Processing Zones." Paper presented at the Annual World Bank Conference on Land and Poverty, Washington, April 8–11.

Moor, Léon de. 1896. "Carte du Congo Belge." Brussels, Belgium: J. Lebègue & Cie. Library of Congress, Geography and Map Division, http://hdl.loc.gov/loc.wdl/dlc.59.

Moran, P. A. P. 1950. "Notes on Continuous Stochastic Phenomena." *Biometrika* 37 (1): 17–23.

Murdock, George P. 1959. *Africa: Its Peoples and Their Culture History.* New York: McGraw-Hill.

Raleigh, Clionadh, Andrew Linke, Håvard Hegre, and Joakim Karlsen. 2010. "Introducing ACLED-Armed Conflict Location and Event Data." *Journal of Peace Research* 47 (5): 1–10.

Schimmer, Russell. 2010. "Congo Free State, 1885–1908." Genocide Studies Program, Yale University, New Haven, CT. http://www.yale.edu/gsp/colonial/belgian_congo/.

Silverman, B. W. 1986. *Density Estimation for Statistics and Data Analysis.* New York: Chapman and Hall.

SRTM 90m (Shuttle Radar Topography Mission, 90 meters) (data set). National Aeronautics and Space Administration (U.S.) (NASA).

Tobler, Waldo. 1993. "Three Presentations on Geographical Analysis and Modeling Non-Isotropic Geographic Modeling Speculations on the Geometry of Geography Global Spatial Analysis." Technical Report 93-1, National Center for Geographic Information and Analysis, Santa Barbara, California.

Uchida, Hirotsugu, and Andrew Nelson. 2009. "Agglomeration Index: Towards a New Measure of Urban Concentration." Background Paper for *World Development Report 2009*, World Bank, Washington, DC.

Vagen, T. G. 2010. "Africa Soil Information Service: Hydrologically Corrected/Adjusted SRTM DEM (AfrHySRTM)." Nairobi, Kenya, and Palisades, NY: International Center for Tropical Agriculture - Tropical Soil Biology and Fertility Institute (CIAT-TSBF), World Agroforestry Centre (ICRAF), Center for International Earth Science Information Network (CIESIN), Columbia University. http://africasoils.net/.

Index

Boxes, figures, maps, notes, and tables are indicated by b, f, m, n, and t, respectively.